New Centrality Measures in Networks

New Centrality Measures in Networks

How to Take into Account the Parameters of the Nodes and Group Influence of Nodes to Nodes

Fuad Aleskerov, Sergey Shvydun
and Natalia Meshcheryakova

CRC Press

Taylor & Francis Group
Boca Raton London New York

CRC Press is an imprint of the
Taylor & Francis Group, an **informa** business
A CHAPMAN & HALL BOOK

First edition published 2022
by CRC Press
6000 Broken Sound Parkway NW, Suite 300, Boca Raton, FL 33487-2742

and by CRC Press
2 Park Square, Milton Park, Abingdon, Oxon, OX14 4RN

© 2022 Fuad Aleskerov, Sergey Shvydun, Natalia Meshcheryakova

CRC Press is an imprint of Taylor & Francis Group, LLC

Library of Congress Cataloging-in-Publication Data

Names: Aleskerov, F. T. (Faud Tagi ogly), author. | Shvydun, Sergey, author. | Mescheryakova, Natalia, author.
Title: New centrality measures in networks : how to take into account the parameters of the nodes and group influence of nodes to nodes
Description: First edition. | Boca Raton, FL : CRC Press, 2022. |
LC record available at https://lccn.loc.gov/2021040809
LC ebook record available at https://lccn.loc.gov/2021040810

ISBN: 9781032063195 (hbk)
ISBN: 9781032066974 (pbk)
ISBN: 9781003203421 (ebk)

DOI: 10.1201/9781003203421

Typeset in Minion
by Deanta Global Publishing Services, Chennai, India

Contents

Acknowledgements

The book has been prepared within the framework of the Basic Research Program at the National Research University Higher School of Economics (HSE) and supported within the framework of a subsidy by the Russian Academic Excellence Project '5–100'.

We would like to thank our colleagues who participated in our joint works – Dr Irina Andrievskaya, Mrs Irina Gavrilenkova, Mrs Alisa Nikitina, Mrs Anna Rezyapova, and Dr Vyacheslav Yakuba.

About the Authors

Fuad Aleskerov graduated from the Mathematical Department of Moscow State University in 1974. He is the Head of the International Center of Decision Choice and Analysis, Head of the Department of Mathematics in the Faculty of Economic Sciences, National Research University Higher School of Economics (HSE University); he also holds the Mark Aizerman Chair on Choice Theory and Decision Analysis, Trapeznikov Institute of Control Sciences of the Russian Academy of Sciences. Aleskerov is on the editorial board for 16 journals, has published ten books and more than 150 articles in peer-reviewed journals, and has presented invited papers at more than 160 international conferences. He is a member of several international scientific societies including Academia Europaea.

Sergey Shvydun earned his PhD degree (*cum laude*) in Applied Mathematics from the National Research University Higher School of Economics (HSE University) in 2020. He is a senior research fellow at HSE's International Center of Decision Choice and Analysis and an associate professor at the Department of Mathematics in the HSE Faculty of Economic Sciences. He is also a senior research fellow at the Laboratory on Choice Theory and Decision Analysis of the Trapeznikov Institute of Control Sciences of the Russian Academy of Sciences. His research interests are in data analysis, social choice theory, and social networks analysis.

Natalia Meshcheryakova is a PhD student in Computer Science at the National Research University Higher School of Economics (HSE University). She received a master's degree in Computer Science from the HSE University in 2018. She is a research fellow at HSE's International Center of Decision Choice and Analysis and a lecturer at the Department of Mathematics in the HSE Faculty of Economic Sciences. She is also a research fellow at the Laboratory on Choice Theory and Decision Analysis of the Trapeznikov Institute of Control Sciences of the Russian Academy of Sciences. Her research interests include social networks, machine learning, and data analysis.

Introduction

T ODAY, NETWORKS ARE USED to represent socio-economic processes, human relations, biological and physical processes, etc. (e.g., Newman 2010, Jackson 2008). Usually, the main, and probably the first, problem studied in networks analysis has been the detection of the most important elements in a network. This very problem was investigated from the first publications on social networks, such as relations between schoolchildren in classes in the 1930s (see Newman 2010). There is also interest nowadays in these topics in Russia (e.g., Gubanov et al. 2011, Gubanov et al. 2019, Kireyev & Leonidov 2018). A very interesting survey was done by Kalinina et al. (2018).

However, in all these models, the vertices in networks are considered to be objects of similar type, i.e., the parameters of the vertices are not taken into account, although in some very important publications this shortage is emphasized openly (e.g., Newman 2010). To attract attention to this problem, let us discuss an example.

Consider a loan of $1 million, borrowed from a bank, and assume that the loan is not repaid in time. If the bank is large it can survive, but for a small, say regional, bank, it might be a cause of bankruptcy. Thus, in the analysis of the loans network, the parameters of the banks should be taken into account.

Let us extend this example to the case of two borrowers who take $500,000 each from the same small regional bank. If one of them repays the loan, the bank will survive, however, if neither borrower returns the loan, it will be the same as the $1 million, and again, the bank will announce bankruptcy.

This last example shows that in the analysis of influence in networks we should take into account the group influence of nodes (players in networks) on an individual node. To the best of our knowledge, this very concept has been discussed in few works (e.g., Myerson 1977).

We can see many examples of this kind. For instance, in the network of international conflicts, the parameters of the countries might be the level of armaments in the countries involved, and the group influence might be evaluated by the military blocs to which the countries belong.

In this book we introduce a class of indices, incorporating these new ideas, and we illustrate the use of these indices via many examples.

THE STRUCTURE OF THE BOOK

In Chapter 1, we introduce the notion of networks, discuss the main classic centrality indices in networks, and introduce new indices – short-range interaction centrality (SRIC) and long-range interaction centrality (LRIC), which differ in the lengths of the path taken into account in the analysis of networks.

In this chapter we also discuss new concepts in the analysis of the power of nodes, based on their interdependence and the impact of indirect connections in network structures. Both these concepts use ideas we have developed before.

Chapter 2 widely illustrates applications of the new indices. It contains the analysis of a global financial market where the countries are key borrowers in the market. Next, we study the networks of international migration, world trade, the global food network, the network of global arms transfers, the network of terrorist groups, and the network of international economic journals. In each case we discuss how to take the parameters of the vertices into account, as well as how to define the group influence in each case.

CHAPTER 1

Centrality Indices in Network Analysis

I N THIS CHAPTER, WE present well-known and widely used classical centrality indices such as different forms of the in- and out-degree indices, centralities based on the eigenvector evaluation, centralities based on the idea of the shortest path, and a centrality index based on cooperative game theory. There are more known indices, and we mention them shortly without presenting their formal definitions. Then, we discuss the shortcomings of classical indices and provide an example showing the necessity of taking into account the parameters of nodes in networks and the possibility of the group influence of nodes to a node in the network. Hence, we propose new classes of indices introduced by our team – short- and long-range interaction centralities (SRIC and LRIC). They take into account not only the features mentioned above but also indirect influence among nodes. Additionally, we have extended the LRIC index for the evaluation of influence in a network where a flow in the network may result in nodes becoming too interdependent on each other, and consequently have some power against each other using the same flow. This measure

DOI: 10.1201/9781003203421-1

is called the interdependence index. Finally, we propose several measures of the edge importance assessment.

1.1 CLASSICAL CENTRALITY MEASURES

Standard methods for the detection of the most influential elements in networks are based on the evaluation of centrality indices for each node, and ranking nodes according to these centrality values (Newman 2010). The higher the centrality of the node, the more important the node is in a network.

There have been many indices developed to measure the centrality level of each node. The measures used have different natures and interpretation and include information about the position of a node and its neighbors in a network. Some of these are based on the number of links to other nodes. Others consider how closely each node is located to other nodes in a network, in terms of distance, or how many times it is on the shortest paths connecting any given pairs of nodes. There are also some indices based on ideas from cooperative game theory and voting theory. In this chapter, we consider the most popular centrality measures known in the literature.

We operate with a graph $G = (V,E)$, where $V = \{1,\ldots,n\}$ is a set of nodes, and $E \subseteq V \times V$ is a set of edges (edge $(i,j) \in E$). We consider undirected and directed graphs. For the latter, the existence of edge (i,j) does not imply the existence of edge (j,i). To describe a graph, we use adjacency matrix $A = [a_{ij}]$, where $a_{ij} = 1$ if there is edge (i,j), and $a_{ij} = 0$ otherwise. Additionally, if connections between nodes are associated with some numerical values, representing the intensity of connections, the graph can be described by a weighted adjacency matrix $W = [w_{ij}]$ that stores the weights of the edges.

1.1.1 Degree Centralities

The simplest centrality measure for undirected graphs is the degree centrality, which is calculated as the total number of neighbors for each node i (Freeman 1979):

$$C_i^{deg} = \sum_{j=1}^n a_{ij} = \sum_{j=1}^n a_{ji} \qquad (1.1)$$

High values of the degree centrality identify nodes with the highest number of connections to other nodes, i.e., nodes for which it is easier to gain access to and/or influence over other nodes locally.

For directed unweighted graphs, four versions of degree centrality measure are possible: in-degree centrality, out-degree centrality, degree centrality, and degree difference. These measures take into account the direction of connections. Additionally, a degree centrality can be adapted to directed or undirected weighted networks. A description of each measure is provided in Table 1.1.

TABLE 1.1 Degree Centrality Measures

Name		Equation	Description
Unweighted graph	In-degree centrality	$C_i^{in-deg} = \sum_{j=1}^n a_{ji}$	The number of incoming edges
	Out-degree centrality	$C_i^{out-deg} = \sum_{j=1}^n a_{ij}$	The number of outgoing edges
	Degree centrality	$C_i^{deg} = \sum_{j=1}^n \left(a_{ij} + a_{ji} \right)$	The total number of i's connections
	Degree difference	$C_i^{diff} = \sum_{j=1}^n \left(a_{ij} - a_{ji} \right)$	The difference between the number of outgoing and incoming edges
Weighted graphs	Weighted in-degree centrality	$C_i^{w\,in-deg} = \sum_{j=1}^n w_{ji}$	The total weight of incoming edges
	Weighted out-degree centrality	$C_i^{w\,out-deg} = \sum_{j=1}^n w_{ij}$	The total weight of outgoing edges
	Weighted degree centrality	$C_i^{w\,deg} = \sum_{j=1}^n \left(w_{ij} + w_{ji} \right)$	The total weight of i's connections
	Weighted degree difference	$C_i^{w\,diff} = \sum_{j=1}^n \left(w_{ij} - w_{ji} \right)$	The difference between the total weight of outgoing and incoming edges

In-degree and out-degree centralities show how a node is affected by its neighbors. For instance, the higher the out-degree centrality of a particular node is, the more nodes are under its control. Degree centrality identifies the most active nodes in different parts of a network, while degree difference is used to evaluate the relative influence of a node on its neighbors. The interpretation of weighted degree centralities is practically the same as for unweighted degree centralities, but weighted measures are more representative than unweighted ones, due to the fact that weighted networks consider the intensities of connections. One should note here that normalized versions of these measures also exist.

1.1.2 Eigenvector Centralities

Since the degree centrality measures do not consider the importance of adjacent nodes, i.e., the information about the degree centrality of its neighbors, several indices have been developed which take this feature into account. An eigenvector centrality considers not only neighboring, but also long-distance connections. The eigenvector centrality (C^{ev}) assigns relative scores to all nodes in a network based on the concept that connections to high-scoring nodes contribute more to the score of the node in question than connections to low-scoring nodes. The idea is that the importance of node i depends on the importance of its neighbors, which, in turn, depends on the importance (degree) of its neighbors, and so on, i.e.,

$$C_i^{ev} = \frac{1}{\lambda} \cdot \sum_{(i,j) \in E} C_j^{ev} = \frac{1}{\lambda} \cdot \sum_{j=1}^{n} C_j^{ev} \cdot A_{ij} \qquad (1.2)$$

The calculation of the centrality measure for each node is related to an eigenvalue problem with respect to adjacency matrix A of a graph: a vector of relative centrality C^{ev} is an eigenvector of the adjacency matrix, i.e.,

$$A \cdot C^{ev} = \lambda \cdot C^{ev}. \qquad (1.3)$$

Generally, all eigenvectors of matrix A can be considered as a centrality measure. However, an eigenvector that corresponds to a maximal eigenvalue is preferable: by the Perron–Frobenious theorem, this vector (and only this, except its co-directional vectors) is positive and real for irreducible non-negative matrix A (Gantmacher 2000) which, by definition, can be presented as a strongly connected graph.

Another generalization of a degree centrality is the Katz centrality (Katz 1953). It measures the weighted count of all paths coming from the node, while a contribution of path of length n is counted with respect to attenuation factor β^n, i.e.,

$$C_i^{Katz} = \beta \sum_{j=1}^{n} a_{ij} + \beta^2 \sum_{j=1}^{n} \left(A^2\right)_{ij} + \ldots = \sum_{k=1}^{\infty} \sum_{j=1}^{n} \beta^k \left(A^k\right)_{ij} \quad (1.4)$$

or in a matrix form

$$C^{Katz} = \left((I - \beta A)^{-1} - I\right) \cdot \vec{e}, \quad (1.5)$$

where $\vec{e}$ is the unit vector, I is the identity vector.

Basically, this measure is applicable to symmetric graphs since the computation of eigenvectors is more difficult for non-symmetric matrices and can produce complex or zero eigenvalues.

Other measures have been introduced to overcome this shortage. An example of these measures is α-centrality (Bonacich 2001). This centrality is defined as the solution of the two-parameter equation

$$C_i^\alpha(\alpha, \beta) = \alpha \cdot \sum_j C_j^\alpha(\alpha, \beta) \cdot A_{ij} + \beta \quad (1.6)$$

or in a matrix form

$$C^\alpha(\alpha, \beta) = (I - \alpha \cdot A)^{-1} \cdot \beta. \quad (1.7)$$

The introduction of parameter β, which corresponds to the initial value of centralities, precludes the possibility of a solution with zero components. In practice, parameter α is selected so that $\alpha < 1/\lambda_{max}$, where λ_{max} is the largest eigenvalue of matrix A. Another example of centrality that can be applied to directed graphs is the PageRank centrality (Brin & Page 1998). According to the model, the importance of a particular node depends on the probability that it be visited by a random walker, i.e.,

$$C_i^{PageRank} = \alpha \cdot \sum_j \frac{C_j^{PageRank}}{C_i^{out-deg}} \cdot a_{ij} + \frac{1-\alpha}{n} \quad (1.8)$$

or in a matrix form

$$C^{PageRank} = \frac{1-\alpha}{n} \cdot \left[I - \alpha \cdot A \cdot \left(I \cdot C^{out-deg} \right)^{-1} \right]^{-1} \cdot \vec{e}, \quad (1.9)$$

where α is the probability of continuing the walk (in general, $\alpha = 0.85$).

Many other measures exist which are based on the idea of eigenvector calculation: Bonacich centrality (Bonacich 1987), hubs and authorities (Kleinberg 1999), subgraph centrality (Estrada & Rodriguez-Velazquez 2005), etc. Note that these measures can be easily adapted to weighted matrix W.

1.1.3 Centralities Based on the Shortest Paths

Another class of centralities is based on the shortest paths between nodes. Two of the most well-known measures are closeness (Bavelas 1950) and betweenness (Freeman 1977) centralities.

Closeness centrality considers how closely each node is located to other nodes of a network in terms of distance. As a result, it elucidates nodes that are the closest to other nodes, and it is usually calculated as

$$C_i^{cl} = \frac{1}{\sum_{j=1}^n d_{ij}}, \quad (1.10)$$

where d_{ij} is the length of the shortest path from node i to node j. A similar idea to closeness centrality was proposed by Rochat (2009), and is called harmonic centrality. This centrality is calculated as the sum of inverse distances between pairs of nodes. If there is no path between a pair of nodes, then the corresponding summand is equal to zero. Thus, this measure performs better on disconnected graphs.

Another way to consider the lengths of the shortest paths is a decay centrality (Jackson 2008). The idea of this measure is to summarize some coefficient $\delta \in (0,1)$ to the power of the lengths of considered paths. A generalized measure of centrality based on closeness was also proposed by Agneessens et al. (2017). The author introduced a tuning parameter δ that measures the importance of geodesic distances and showed that using the parameter, degree-centrality and closeness centrality are two specific instances of their more general measure.

Betweenness centrality detects nodes that lie on the shortest paths between any other two nodes most of the time. It is defined as

$$C_i^{btw} = \sum_{jk} \frac{\sigma_{jk}(i)}{\sigma_{jk}}, \tag{1.11}$$

where σ_{jk} is the number of the shortest paths from node j to node k, $\sigma_{jk}(i)$ is the number of the shortest paths from node j to node k going through the node i. High betweenness centrality identifies nodes that are crucial hubs and/or bridges between disparate clusters in a network.

A measure similar to betweenness centrality is stress centrality (Shimbel 1953). The main difference is that coefficients of stress centrality are not normalized to the total number of the shortest paths between considered nodes. The concept of betweenness centrality was extended to a group level in Everett & Bogatti (1999). There is also percolation centrality (Piraveenan et al. 2013) which is based on the idea that each considered path is weighted to the contribution of this path to a percolation process, while the weights of paths depend on the percolation level of a source node and the total percolation state of a network. Contrary to the centralities that are described above, percolation centrality requires initial conditions of the level of percolation of each node.

1.1.4 Centralities from Cooperative Game Theory

An influence in networks is also evaluated in the field of game theory and mechanism design. There are various power indices that are applied to the network theory. In this case, a network is interpreted as a set of interacting individuals that contribute to a total productive value of a network, and the problem is how to share this generated value among them.

Myerson (1977) proposed a value that is based on the Shapley–Shubik index (Shapley & Shubik 1954). The Myerson value shows an average contribution for each node, where the contribution is a function v generated by the network, with and without this individual, i.e.,

$$ C_i^{MV}(G,v) = \sum_{S \in V} \frac{(|S|-1)!(|N|-|S|)!}{|N|!} \left(v(S) - v(S \setminus \{i\}) \right), \quad (1.12) $$

The main disadvantages of this approach are its large computational complexity (since it requires consideration of all possible subgraphs) and uncertainty (about how the value of a subgraph should be assigned). Partial rankings of nodes based on the neighborhood-inclusion principle were also discussed by Schoch (2018).

Many other approaches to key nodes detection exist. For instance, a different perspective of the central nodes' estimation was proposed by Kang et al. (2012) which is called diffusion centrality. This centrality takes into account the attributes of nodes and their properties, and measures the quality of diffusion of a property p starting from a node v. Despite the completeness of the proposed measure, some prior conditions should be known as conditional probabilities of the 'infection' or so-called 'diffusion rules'.

Many real networks are complex, and their elements are not homogenous. The nodes of a network may have various individual attributes that characterize their size, importance, level of influence, etc. This possibility was mentioned by Newman (2003): '[...] and vertices or edges may have a variety of properties, numerical or otherwise, associated with them'. For instance, the threshold of influence, which indicates the level when this node becomes affected, may give the result that even connections with the same weight w can be influential for node i, and not influential for some other node, j, depending on the attributes of these nodes. The size or importance of nodes may lead to the situation that influence on a group of nodes may, in total, contribute less than influence on a single node. Finally, some nodes can influence other nodes only in collaboration with some other members.

All these aspects show that power distribution in networks should be evaluated with respect to both the individual attributes of nodes and the connections between them. Unfortunately, centrality measures based on degree or the shortest paths do not take into account the nodes' features, and consequently cannot be applied since initial connections of a network do not represent the actual picture of the nodes' influence. Diffusion centrality considers particular attributes of nodes, but it lacks such features as individual thresholds of influence, nodes' importance, and the possibility of group influence. Moreover, the way to define diffusion rules in various applications is unclear. On the other hand, the influence measure proposed by Myerson (1977) considers group influence but does not take into consideration the individual characteristics of nodes, or the possibility of indirect influence. Therefore, we consider new centrality measures that consider the individual attributes of nodes, as well as their group and indirect influence.

1.2 SHORT- AND LONG-RANGE INTERACTION CENTRALITY INDICES

1.2.1 Individual and Group Influence

First, we introduce parameter q_j that each node can possess in an explicit form. This possibility was mentioned by Newman (2003).

Assume parameter q_j is the threshold of influence, which indicates the level when this node becomes affected. This parameter may depend on some external parameters of a node or may be calculated with respect to a network structure and a node position in a network.

The threshold of influence appears in many applications. For instance, in computer networks, nodes have a finite buffer size, a limit on the amount of traffic (i.e., throughput) that they can handle. In financial networks, banks have a minimum required level of liquidity to remain stable during stress. In trade networks, countries have requirements for the amount of food needed to support minimum per capita nutritional standards. In mobility networks, nodes have a limited capacity for storing and processing the agents. These threshold models also appear in epidemic spreading, and in the analysis of users' opinions.

Another important motivation of our work is that nodes might influence each other not separately, but in groups. For instance, consider a small banking network with three borrowers and one lender (Figure 1.1):

Let the total assets of Bank 4 (that is a lender in the network) equal US$20. Assume that the lender provides a loan to three other banks (Banks 1, 2, and 3) and it suffers serious damage when the borrowers do not repay 50% of its total assets (hence, $q_4 = 20 \cdot 50\% = 10$). In that case, borrowers 1, 2, and 3, with the

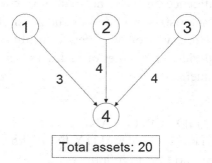

FIGURE 1.1 Small banking network.

loans $w_{41} = 3$, $w_{42} = 4$, and $w_{43} = 4$ do not create a problem for Bank 4 separately. However, taken together, they might represent a serious risk to the lender, since $w_{41} + w_{42} + w_{43} > q_4$.

Obviously, borrowers do not always influence lenders individually or in groups. In fact, this very example provides a more detailed understanding of the threshold value. If the total assets of Bank 4 were high, then none of the borrowers could create a serious problem for the bank. Thus, we provide definitions of critical groups and their pivotal members.

Definition 1. A critical group $\Omega(j)$ for node j is a subset of nodes whose total influence exceeds the value of the parameter prescribed to vertex j, which, for short, we call the threshold q_j. More formally, $\Omega(j) \subseteq V$ is a critical group for node j, if

$$\sum_{i \in \Omega(j)} w_{ij} \geq q_j. \tag{1.13}$$

Definition 2. A node is pivotal for a group if its exclusion from this group makes the group non-critical. There can be several pivotal nodes in one group.

Formally, $\Omega^p(j) \subseteq \Omega(j)$ is a subset of pivotal nodes of group $\Omega(j)$ if

$$\forall k \in \Omega^p(j) \qquad \sum_{i \in \Omega(j) \setminus \{k\}} w_{ij} < q_j. \tag{1.14}$$

To illustrate the idea of Definitions 1 and 2, consider the example from Figure 1.1. Assume that the threshold value for node 4 is 8, i.e., $q_4 = 8$. In this case, there are two critical groups for node 4: $\Omega_1(4) = \{2,3\}$ (total influence is 8), $\Omega_2(4) = \{1,2,3\}$ (total influence is 11). Nodes 2 and 3 are pivotal for these groups. However, node 1 is not pivotal in group $\Omega_2(4)$ because after its exclusion, the group $\{2,3\}$ is still critical.

Moreover, nodes can influence other nodes indirectly. Thus, we introduce a definition of a simple path to identify all channels of indirect influence.

Definition 3. A simple path between nodes i and j in graph G is a sequence of edges that connects them and contains distinct nodes. More precisely, (i,k_1), (k_1, k_2), (k_2, k_3),..., (k_{s-1},j) such that I $\neq k_1 \neq \cdots \neq k_{s-1} \neq j$ is a simple path of length s denoted as $P^t_{i-j}(s)$, where t is the number of the path.

The definition of simple paths is illustrated in Figure 1.2. There are three simple paths from node 2 to node 3: $P^1_{2-3}(2) = \{(2,1), (1,3)\}$, $P^2_{2-3}(3) = \{(2,5), (5,1), (1,3)\}$, $P^3_{2-3}(5) = \{(2,6), (6,7), (7,8), (8,1), (1,3)\}$. Each of these paths contains distinct nodes. However, a sequence $(2,5)$, $(5,1)$, $(1,7)$, $(7,8)$, $(8,1)$, $(1,3)$ is not a simple path, because node 1 appears twice in the sequence.

1.2.2 Short-Range Interaction Centrality (SRIC)

Now, we describe a power index that was first introduced in the voting theory (Aleskerov 2006) and was then adapted to estimate the influence of agents in the loan market network (Aleskerov et al. 2014). This is called short-range interaction centrality (SRIC), and this index takes into account the influence of a node within groups, of one mediate node in the estimation of indirect

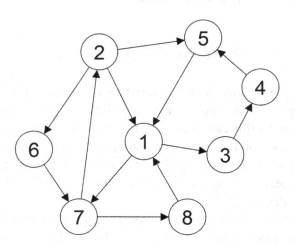

FIGURE 1.2 Simple paths between nodes 2 and 3.

influence (that is why the index is called short range), and it also includes an individual characteristic of a node – a threshold.

To illustrate the idea of the model, we use the definitions of critical groups and pivotal nodes introduced in Section 1.2.1, and consider the graph presented in Figure 1.3. In constructing such a graph, we investigate how each node influences other nodes in a network.

The construction of the short-range interactions matrix includes several steps. First, we need to find all possible critical groups for a specific node. Second, we need to estimate the contribution of each pivotal node from the critical groups, based on normalized intensities of direct and indirect influence. The final step is a normalization of the vectors obtained.

For example, consider the influence of the nodes 2, 3, 4, and 5, on node 1. Since the threshold of the node 1 is 4 ($q_1 = 4$), we can find all critical groups and their pivotal members for this node. Notice that critical groups may include not only direct neighbors, but also indirectly connected nodes. We assume that

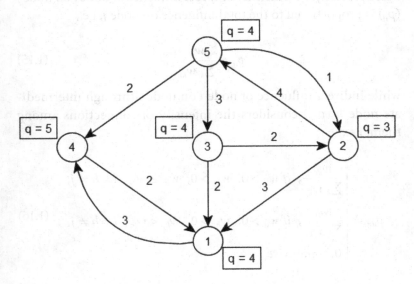

FIGURE 1.3 Numerical example.

TABLE 1.2   List of Critical Groups and Pivotal Nodes
in These Groups for Node 1

Number, k	Critical Groups, $\Omega_k(1)$	Pivotal Nodes, $\Omega^P_k(1)$
1	$\Omega_1(1) = \{2,3\}$	$\Omega^P_1(1) = \{2,3\}$
2	$\Omega_2(1) = \{2,4\}$	$\Omega^P_2(1) = \{2,4\}$
3	$\Omega_3(1) = \{3,4\}$	$\Omega^P_3(1) = \{3,4\}$
4	$\Omega_4(1) = \{2,3,4\}$	$\Omega^P_4(1) = \emptyset$
5	$\Omega_5(1) = \{2,3,5\}$	$\Omega^P_5(1) = \{2,3\}$
6	$\Omega_6(1) = \{2,4,5\}$	$\Omega^P_6(1) = \{2,4\}$
7	$\Omega_7(1) = \{3,4,5\}$	$\Omega^P_7(1) = \{3,4\}$
8	$\Omega_8(1) = \{2,3,4,5\}$	$\Omega^P_8(1) = \emptyset$

indirect neighbors add zero value into a group weight, and they
never become pivotal. Critical groups and pivotal nodes for these
groups are represented in Table 1.2.

Then we calculate normalized intensities of direct and indirect influence through one intermediate node. Note that indirect
influence is calculated only if a direct link with node 1 exists.
More precisely, the intensity of direct influence of node i on node j
(p_{ij}) is proportional to the total influence on node j, i.e.,

$$p_{ij} = \frac{w_{ij}}{\sum_k w_{kj}}, \qquad (1.15)$$

while indirect influence of node i on node j through intermediate node h (p_{ihj}) considers the intensity of connections among
them, i.e.,

$$p_{ihj} = \begin{cases} \dfrac{w_{ih}}{\sum_k w_{kj}}, & \text{if } w_{ij} > 0, \ w_{hj} > 0, \ w_{hj} \geq w_{ih}, \ i \neq h \neq j, \\[2ex] \dfrac{w_{hj}}{\sum_k w_{kj}}, & \text{if } w_{ij} > 0, \ w_{hj} > 0, \ w_{hj} < w_{ih}, \ i \neq h \neq j, \\[2ex] 0, & \text{otherwise} \end{cases} \qquad (1.16)$$

Indirect intensity has the following interpretation. If node i is not directly connected to node j, the intensity of its influence is equal to zero. However, if node i is directly connected to the other neighbor of node j (node h), the intensity of indirect influence of node i on node j through node h should not be greater than the intensity of direct influence of node h on node j.

For the numerical example, the direct intensities of influences on node 1 are $p_{21} = 3/7$; $p_{31} = 2/7$; $p_{41} = 2/7$, and the indirect intensities of influences with node 1 are $p_{321} = 2/7$. Note that node 5 is not directly connected to node 1, thus, its intensity of influence is equal to zero. Direct and indirect intensities of influence on node 1 are provided in Table 1.3.

Knowing critical groups, pivotal nodes, and normalized intensities of connections, we calculate the intensities within every critical group for each pivotal node there. We also pay attention to the size of a group (the bigger a group is, the less contribution each pivotal node has). Hence, we introduce the intensity of the influence of pivotal node i on j within group $\Omega_k(j)$, denoted by $f(i,\Omega_k(j))$, i.e.,

$$f\left(i,\Omega_k(j)\right) = \frac{p_{ij} + \sum_{h \in \Omega_k(j)} p_{ihj}}{\left|\Omega_k(j)\right|}, \quad i \in \Omega_k^p(j). \tag{1.17}$$

For our numerical example, the results of these intensities are represented in Table 1.4.

After we get intensities of influence of node i on node j in all critical groups where i is pivotal, we need to aggregate them into

TABLE 1.3 Direct and Indirect Intensities of Influence on Node 1

Nodes	2	3	4	5
2	3/7	0	0	0
3	2/7	2/7	0	0
4	0	0	2/7	0
5	0	0	0	0

TABLE 1.4　Intensity of an Influence on Node 1 within Group $\Omega_k(j)$

k	Critical Groups, $\Omega_k(1)$	Pivotal Nodes, $\Omega^p{}_k(1)$	Influence of Pivotal Nodes on Node j in Critical Group
1	$\Omega_1(1) = \{2,3\}$	$\Omega^p{}_1(1) = \{2,3\}$	$f(2,\Omega_1(1)) = 3/14,$ $f(3,\Omega_1(1)) = 4/14.$
2	$\Omega_2(1) = \{2,4\}$	$\Omega^p{}_2(1) = \{2,4\}$	$f(2,\Omega_2(1)) = 3/14,$ $f(3,\Omega_2(1)) = 2/14.$
3	$\Omega_3(1) = \{3,4\}$	$\Omega^p{}_3(1) = \{3,4\}$	$f(3,\Omega_3(1)) = 2/14,$ $f(4,\Omega_3(1)) = 2/14.$
4	$\Omega_4(1) = \{2,3,4\}$	$\Omega^p{}_4(1) = \emptyset$	-
5	$\Omega_5(1) = \{2,3,5\}$	$\Omega^p{}_5(1) = \{2,3\}$	$f(2,\Omega_5(1)) = 3/21,$ $f(3,\Omega_5(1)) = 4/21.$
6	$\Omega_6(1) = \{2,4,5\}$	$\Omega^p{}_6(1) = \{2,4\}$	$f(2,\Omega_6(1)) = 3/21,$ $f(4,\Omega_6(1)) = 2/21.$
7	$\Omega_7(1) = \{3,4,5\}$	$\Omega^p{}_7(1) = \{3,4\}$	$f(3,\Omega_7(1)) = 2/21,$ $f(4,\Omega_7(1)) = 2/21.$
8	$\Omega_8(1) = \{2,3,4,5\}$	$\Omega^p{}_8(1) = \emptyset$	-

a single value, or, in other words, node-to-node influence. The aggregation proposed by Aleskerov et al. (2014) is as follows:

$$\chi_i(j) = \Sigma_k f\left(i,\Omega_k(j)\right). \tag{1.18}$$

Finally, these values are normalized to their sum

$$\hat{\chi}_i(j) = \frac{\chi_i(j)}{\Sigma_k \chi_k(j)}. \tag{1.19}$$

For our example, the result is

$$\chi_2(1) = \frac{3}{14} + \frac{3}{14} + \frac{3}{21} + \frac{3}{21} = \frac{15}{21};$$

$$\chi_3(1) = \frac{4}{14} + \frac{2}{14} + \frac{4}{21} + \frac{2}{21} = \frac{15}{21};$$

$$\chi_4(1) = \frac{2}{14} + \frac{2}{14} + \frac{2}{21} + \frac{2}{21} = \frac{10}{21};$$

$$\chi_5(1) = 0.$$

The normalized values are

$$\hat{\chi}_2(1) = \frac{15}{40}; \quad \hat{\chi}_3(1) = \frac{15}{40}; \quad \hat{\chi}_4(1) = \frac{10}{40}; \quad \hat{\chi}_5(1) = 0.$$

Derived values can be treated as new weights on edges between ith node and jth node: this is a final modified influence of node i on node j that includes both direct and indirect influence, individually or via groups with other nodes, taking into account individual parameters of nodes. For our example, we can conclude that nodes 2 and 3 influence node 1 the most, and node 5 does not influence node 1 at all. Similarly, we can evaluate the final influence corresponding to this approach for each other node and obtain the matrix of node-to-node influence (see Table 1.5).

Note that although node 5 is connected to node 3, it does not influence it, according to the predefined thresholds. This feature is not taken into consideration by classical centrality measures. For instance, node 5 is the second most influential node according to the weighted out-degree centrality, and the most influential by weighted degree centrality. Moreover, node 5 is ranked higher than node 3 by eigenvector and PageRank centralities, even though node 3 influences node 5 via node 2.

Information about node-to-node influence can be converted into a single value that corresponds to each node (influence measure in a graph) – SRIC index α_{SRIC} which is introduced in

TABLE 1.5   Intensity of an Influence on Node 1 within Group $\Omega_k(j)$

Nodes	1	2	3	4	5
1	0	0	0	0.6	0
2	0.375	0	0	0	1
3	0.375	0.667	0	0	0
4	0.25	0	0	0	0
5	0	0.333	0	0.4	0

Aleskerov et al. (2014). Convolution of node-to-node influence may be implemented with the help of different techniques that may require externally introduced weights of nodes, or a node structure within a graph. For example, the index may be obtained as a weighted sum of the node-to-node influences where weights are represented as normalized out-degrees. The idea is that the influence on the particular node is more valuable if this node controls many other nodes. Another option of the influence aggregation is a mean value for each node. More formally,

$$C_i^{SRIC} = \frac{\sum_{j \in V} \hat{\chi}_i(j)}{|V|}. \tag{1.20}$$

Note that the denominator is the same for each node i. If we omit it, the rank of nodes remains the same. For the numerical example, the aggregated power by the mean value gives the following results:

$$C_1^{SRIC} = 0.12; \ C_2^{SRIC} = 0.28; \ C_3^{SRIC} = 0.21;$$

$$C_4^{SRIC} = 0.05; \ C_5^{SRIC} = 0.15.$$

The SRIC index definitely provides influence estimation in a graph, but it also has some drawbacks. First, SRIC requires the existence of a direct connection for a possibility of indirect influence which usually leads to underestimation of the real influence. Indeed, in our example, node 5 may influence node 1 through node 4, but SRIC is not sensitive to this connection. Second, the index considers only short connections and does not take into account the influence through more than one intermediate node. Thus, there is a need to consider indirect connections between nodes. Third, group formation procedure includes indirectly connected nodes, and it appears unreasonable as these nodes do not influence the one under consideration at all. Next, intensities of indirect influence appear controversial, as intermediate nodes may have their own thresholds of influence. Moreover, the algorithm requires the enumeration of all subsets of a nodes' set, which is a computationally complex task. However, since the idea of group influence, which depends on individual parameters of nodes, corresponds to many real-life problems, we propose several more general models based on this approach.

Remark. As another method of aggregation, we can also calculate classic centrality measures (e.g., weighted degree centralities or PageRank centrality) on a new graph with normalized influences, obtained in Table 1.5.

1.2.3 Long-Range Interaction Centrality (LRIC)

Let us consider new models with many intermediate nodes when we estimate the influence between two nodes which are not necessarily directly connected. As in Aleskerov et al. (2014), our methodology is based on a group influence, but the procedure of group formation is different. We also include many options for additional parameters (threshold formation, aggregation procedures, etc.).

The first step requires the rewriting of given weights on edges with intensities of influence. In other words, we calculate direct influences on this stage. Here we also assume that each node has a predefined threshold (given or calculated based on a graph), which defines the level of influence. For SRIC index intensities are calculated through critical groups. However, in our approach, we take into account groups of direct neighbors only. When we calculate the influence of node i on node j we form all critical groups where i is pivotal, and consider the one in which i directly influences j most. Formally,

$$c_{ij} = \begin{cases} \max\limits_{\Omega_k(j):i\in\Omega_k^p(j)} \dfrac{w_{ij}}{\sum_{h\in\Omega_k(j)} w_{hj}}, & \textit{if } \exists \Omega_k^p(j):i\in\Omega_k^p(j), \\ 0, & \text{otherwise.} \end{cases} \quad (1.21)$$

According to Equation (1.21), $0 \leq c_{ij} \leq 1$; $c_{ij} = 0$ implies either i does not have a direct connection to j, or the weight on edge (i,j) is too small and i is not a pivotal node in any critical groups of j. Hence, some weak arrows may disappear compared to the original graph. Thus, $c_{ij} = 1$ implies that node i individually may exceed the threshold of j.

To illustrate the idea of direct influence we consider the example which was discussed for the SRIC index (see Figure 1.4).

For the numerical example, critical groups and the direct influence of their members, are provided in Table 1.6.

The new graph of direct influences according to Equation (1.21) is provided in Figure 1.5.

Obviously, one node may also influence another node indirectly. Thus, it is also necessary to consider indirect paths among nodes. A long-range interactions approach may be subject to different contexts and ideas. In current work we propose a path-based technique. We use three definitions given in Section 1.2.1

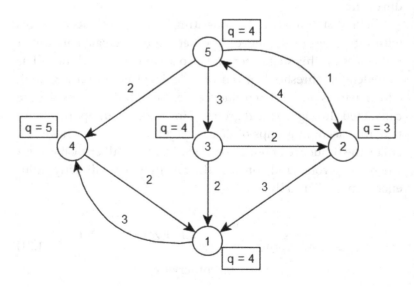

FIGURE 1.4   Numerical example.

TABLE 1.6   Direct Influence Calculation

Node j	Adjacent Nodes, i	Critical Group Where i Influences j the Most	Weight of the Group	Direct Influence, c_{ij}
1	2	$\Omega(1) = \{2,3\}$ or $\{2,4\}$	5	$c_{21} = 3/5$
	3	$\Omega(1) = \{3,4\}$	4	$c_{31} = 2/4$
	4	$\Omega(1) = \{3,4\}$	4	$c_{41} = 2/4$
2	3	$\Omega(2) = \{3,5\}$	3	$c_{32} = 2/3$
	5	$\Omega(2) = \{3,5\}$	3	$c_{52} = 1/3$
3	5	$\varnothing$	-	$c_{53} = 0$
4	1	$\Omega(4) = \{1,5\}$	5	$c_{14} = 3/5$
	5	$\Omega(4) = \{1,5\}$	5	$c_{54} = 2/5$
5	2	$\Omega(5) = \{2\}$	4	$c_{54} = 1/1$

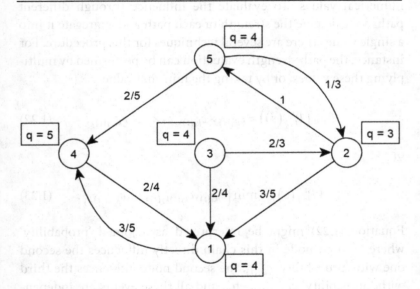

FIGURE 1.5   Direct influences for the numerical example.

– critical groups (Definition 1), pivotal nodes (Definition 2), and a simple path (Definition 3).

Thus, to construct a matrix of total influence, we consider a graph of direct influences and analyze all simple paths between nodes (Definition 3).

TABLE 1.7   Simple Paths between Nodes 3 and 1

t, Path ID	s, Path Length	$P_{3-1}^t(s)$
1	1	$\overset{2/4}{3\to1}$
2	2	$\overset{2/3}{3}\overset{3/5}{\to2\to1}$
3	4	$\overset{2/3}{3}\overset{1}{\to2}\overset{2/5}{\to5}\overset{2/4}{\to4\to1}$

To illustrate the idea, we evaluate the influence of node 3 on node 1. There are three simple paths from node 3 to node 1, enumerated in Table 1.7.

As we can see above, each path is characterized by a set of numerical values. To evaluate the influence through different paths we calculate the strength of each path and aggregate it into a single value. There are several techniques for this procedure. For instance, the path strength evaluation can be performed by multiplying these values, or by taking the minimal value

$$f_{mult}\left(P_{i-j}^t(s)\right) = c_{ik_1(t)} \times c_{k_1(t)k_2(t)} \times \ldots \times c_{k_{s-1}(t)j}, \qquad (1.22)$$

or

$$f_{min}\left(P_{i-j}^t(s)\right) = \min\left\{c_{ik_1(t)}, c_{k_1(t)k_2(t)}, \ldots, c_{k_{s-1}(t)j}\right\}. \qquad (1.23)$$

Equation (1.22) might be interpreted as a joined 'probability' where the first node in this chain directly influences the second one with 'probability' $c_{ik_1(t)}$, the second node influences the third with 'probability' $c_{k_1(t)k_2(t)}$, etc., and all these events are independent. The main advantage of this equation is that indirect influence decreases with distance. Equation (1.23) may be thought of as flow capacity, when node i cannot influence node j through path P_{i-j}^t more than the minimal influence on this path.

As we study the influence of node 3 on node 1, let us enumerate all simple paths between these nodes once again, calculate f_{mult}

and f_{min} for each of them, and estimate the influence as the maximal path strength (see Table 1.8).

As there are many paths between nodes, we need to aggregate the pairwise influence into a single value. The choice of the rules depends on the problem, since they take into account different features. For instance, the total influence between two nodes can be calculated as the value of maximal path strength, or as the value of accumulated paths strength, i.e.,

Accumulated paths strength: $c_{ij}^{sum} = \min\left(\Sigma_k\, f\left(P_{i-j}^k(s)\right),1\right)$;

Maximal path strength: $c_{ij}^{max} = \max_k\, f\left(P_{i-j}^k(s)\right)$.

For most of the real networks, very long-range interactions have no meaning. Taking into account this fact, we can limit the length of considered paths with a parameter, s, and do not study paths of length of more than s. Practically, this parameter is usually not longer than 4 (in all examples we have considered that larger numbers do not change the results drastically).

Since there are various models of indirect paths evaluation, we can propose different versions of the LRIC measure (see Table 1.9). Note that due to poor interpretation, we do not consider the version of the LRIC model that evaluates each path using Equation (1.23) and then accumulates all of them.

Thus, we can estimate the influence between each of the other pairs of nodes. For the numerical example, the node-to-node

TABLE 1.8   Indirect Influence of Node 3 on Node 1 through Different Paths

t, Path ID	P_{3-1}^t	$f_{mult}\left(P_{3-1}^t\right)$	$f_{min}\left(P_{3-1}^t\right)$
1	$3 \xrightarrow{2/4} 1$	2/4	2/4
2	$3 \xrightarrow{2/3} 2 \xrightarrow{3/5} 1$	$2/3 \times 3/5 = 2/5$	3/5
3	$3 \xrightarrow{2/3} 2 \xrightarrow{1} 5 \xrightarrow{2/5} 4 \xrightarrow{2/4} 1$	$2/3 \times 1 \times 2/5 \times 2/4 = 2/15$	2/5

TABLE 1.9   Versions of LRIC Measure

Paths Influence\ Paths Aggregation	Accumulated Paths Strength	Maximal Path Strength
Joined 'probability' (f_{mult})	LRIC (Sum)	LRIC (Max)
Flow capacity (f_{min})	-	LRIC (MaxMin)

TABLE 1.10   Node-to-Node Influence for the Numerical Example

Nodes	1	2	3	4	5
1	0	0	0	0.6	0
2	0.6	0	0	0.6	1
3	0.5	0.667	0	0.6	0.667
4	0.5	0	0	0	0
5	0.4	0.333	0	0.4	0

influence, which is calculated through identification of maximal path influence, which is denoted by LRIC (Max), is provided in Table 1.10.

We have obtained one version of the matrix of long-range interaction which contains information about pairwise influence. As in the short-range approach, we can apply different aggregation rules to obtain the nodes' influence overall. For instance, according to normalized mean value convolution, the C_i^{LRIC} index gives the following results: $C_1^{LRIC} = 0.09$; $C_2^{LRIC} = 0.32$; $C_3^{LRIC} = 0.36$; $C_4^{LRIC} = 0.07$; $C_5^{LRIC} = 0.16$. The most important elements are nodes 3 and 2.

Thus, we described a novel method for the influence estimation in networks which transforms the initial network to the network of influence with respect to individual attributes of nodes, and their group influence, and then aggregates this information into SRIC and LRIC indices. It is necessary to mention that the proposed approach also provides information about node-to-node influence in contrast to classical centrality measures. Information about the dependency relationship between nodes allows us to identify the most vulnerable participants, and also makes it easier to interpret different processes in a network.

The original work for the SRIC (called a key borrower index) and LRIC indices are presented in Aleskerov et al. (2014, 2016a, 2016b). One should also note that in Meshcheryakova & Shvydun (2018), we proposed a novel method for the influence estimation in networks which transforms the initial network to the network of influence and focuses on the analysis of possible chain reactions in the obtained network (the so-called domino or contagion effect) in the form of simulations.

1.3 POWER OF NODES BASED ON THEIR INTERDEPENDENCE

Central elements in networks are important to understand the real world systems which we try to model. However, connections in a network may have several meanings which create some difficulties in detecting such elements. As an example, we can consider citation networks in which the edge from the vertex A to the vertex B means that B influences A. As another example, we can consider a flow network where the edge from the vertex A to the vertex B means that A influences B.

In some cases, nodes might influence each other via the same edge. For instance, trade relations make countries increasingly economically interdependent and highly integrated. Indeed, importers depend on exporters for the purchase of products not manufactured domestically. The increase of trade relations and the specialization of countries in the production of particular goods leads to the situation that exporters may use trade to achieve political goals. The power of a country increases with the volume of its exports. However, importers can gain certain power over exporters using trade relations as a potential leverage (Mancheri 2015). For instance, an importer might suddenly refuse the purchase of imported goods, which might be considered as a threat for political concession. Since a potential import ban may lead to economic losses for the exporter, it might affect the economic growth of the exporter, its employment, and its financial stability. The importer also influences the exporter, as the exporter seeks the importer's market and investments. Therefore, the edge from node A to node B means that both nodes are highly interdependent.

Consider a small network in Figure 1.6, and suppose that the connections correspond to trade flows.

According to Figure 1.6(a), node A is a major exporter of goods for node B, thus it influences node B. On the other hand, node B is the largest importer of goods from A, hence, node A may also depend on node B. Therefore, nodes A and B can be considered to be interdependent.

Consider another example. If node B is not the largest importer for node A (see Figure 1.6(b)), then the influence of node B to node A can be evaluated as insignificant. Similarly, if node B is the major importer for node A, while node A is not the major exporter for node B (see Figure 1.6(c)), there is only a one-sided influence between nodes (node B influences node A).

Finally, if these nodes are not the largest exporters and importers for each other (see Figure 1.6(d)), then we assume that there is no nodal interdependence between A and B.

Although influence on Figure 1.6 can be evaluated using existing centrality concepts by considering separately how nodes influence each other as exporters and as importers (one-sided nodes

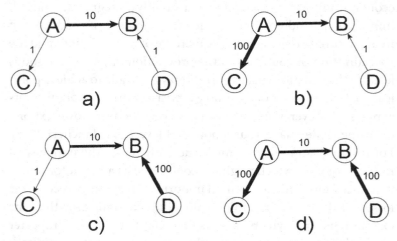

FIGURE 1.6   Interdependence of nodes in networks.

influence) – we did not find any measure that takes into account all the cases from Figure 1.6 simultaneously. The most relevant measure is the hyperlink-induced topic search (HITS) algorithm introduced by Kleinberg (1999). The algorithm assigns two scores – the hub score and the authority score – and, according to the model, the node has a high hub score if it is connected to high authority nodes. Similarly, a node possesses a high authority score if it is pointed by nodes with high hub scores. However, it can be easily seen that the HITS algorithm does not capture the nodes' interdependence. Moreover, the HITS algorithm does not consider a heterogeneity of nodes or their potential group influence on other nodes. Our approach takes into account the main features of interdependence, which we discussed using Figure 1.6.

To evaluate the importance of an edge (i,j) we extend the idea of individual and group influence which is discussed in Section 1.2.1. We assume that each node has two parameters q_i^{in} and q_i^{out}, which indicate the level when node i becomes affected. Similarly, to the SRIC and LRIC indices, q_i^{in} corresponds to a threshold of influence with respect to internal links. In other words, if the total influence of nodes that are connected to node i exceeds the threshold q_i^{in}, they will influence node i. On the contrary, q_i^{out} corresponds to the importance of out-going links from node i. For instance, in trade networks, as node i also benefits from connections to other nodes, q_i^{out} corresponds to the minimal export value which will be critical (e.g., economically) for node i to lose. As was discussed above, q_i^{in} and q_i^{out} can be defined using the network structure or externally.

We can adapt the idea of direct influence evaluation using a multilayer network concept from Section 1.2.1 to evaluate c_{ij}^{in} and c_{ij}^{out} representing the importance of flow w_{ij} for nodes i and j. Formally,

$$c_{ij}^{in} = \begin{cases} \max_{\Omega_k(j): i \in \Omega_k^{p,in}(j)} \dfrac{w_{ij}}{\sum_{h \in \Omega_k(i)} w_{hj}}, & \text{if } \exists \Omega_k^{p,in}(j) : i \in \Omega_k^{p,in}(j), \\ 0, & \text{otherwise,} \end{cases} \tag{1.24}$$

$$c_{ij}^{out} = \begin{cases} \max\limits_{\Omega_k(i):i\in\Omega_k^{p,out}(i)} \dfrac{w_{ij}}{\sum_{h\in\Omega_k(i)} w_{ih}}, & \text{if } \exists\Omega_k^{p,out}(i): j\in\Omega_k^{p,out}(i), \\ 0, & \text{otherwise}, \end{cases} \quad (1.25)$$

where $\Omega_k^{p,in}(j)$ is a critical group for node j with respect to its threshold value q_j^{in} and $\Omega_k^{p,out}(i)$ is a critical group for node i with respect to its threshold value q_i^{out}.

Thus, we are able to evaluate the importance of flow w_{ij} for both nodes i and j. Then such information can be expressed using two matrices of pairwise influence $C^{in} = [c_{ij}^{in}]_{n\times n}$ (influence level) and $C^{out} = [c_{ij}^{out}]_{n\times n}$ (dependence level). One can represent such information as a multiplex network that consists of two layers: the layer of nodes' influence and the layer of nodes' dependence (see Figure 1.7).

We defined above how nodes influence each other through the adjacent edge. If $c_{ij}^{in} \gg c_{ij}^{out}$, then node i influences node j, but not vice versa. Similarly, if $c_{ij}^{in} \ll c_{ij}^{out}$ then node i is highly dependent on node j, while node j is not directly influenced by node i. Finally, if $c_{ij}^{in} \approx c_{ij}^{out}$, both nodes are dependent on each other. Then we can use this information to define the level of direct interdependence of the nodes.

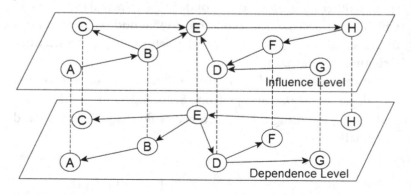

FIGURE 1.7 Multiplex representation of nodes' interdependence.

However, these evaluations do not allow us to take into account the possibility of indirect influence of nodes, since nodes can affect each other via some intermediate vertices. Hence, we need to identify how nodes might be interdependent indirectly.

In this way, we construct several models of nodes' interdependence with respect to the network structure.

1) Model 1: nodes' interdependence as aggregation of indirect influences.

Since nodes influence each other indirectly, we can extend matrices of pairwise influence C^{in} and C^{out} to matrices of indirect influence. This step can be taken considering different paths in a graph of direct influence (see Section 1.2.3). Hence, one can construct the matrices $\hat{C}^{in}$ and $\hat{C}^{out}$ corresponding to the level of indirect influence/dependence of nodes. As we already know, $\hat{c}_{ij}^{in}$ defines how node i influences node j, directly or indirectly, and $\hat{c}_{ij}^{out}$ evaluates how node i is dependent on node j. Then we aggregate this information into a single matrix $\hat{C}$. For instance, $\hat{c}_{ij} = \hat{c}_{ij}^{in} - \hat{c}_{ij}^{out}$ can be used as an interdependence value. If $\hat{c}_{ij} \approx 0$, then the nodes do not influence each other. If $\hat{c}_{ij} > 0$ node i has more influence on node j. On the other hand, if $\hat{c}_{ij} < 0$, node i is dependent on node j.

Information about an aggregated node-to-node interdependence level can be converted into a single index $\hat{c}_i$ that corresponds to the overall level of influence in a network. For example, the index for node i may be obtained as a normalized total influence of node i on other nodes in a network.

2) Model 2: nodes' interdependence based on difference of direct influences.

Again, we construct another model to evaluate the indirect influence of nodes. However, contrary to the previous model, we first analyze how nodes influence each other in a direct way. For example, if we consider trade networks, and $c_{ij}^{in} = 1$, node i is the major exporter of goods for node j. Hence, it has high influence on node j if it uses the shortage of the export as a political tool. However, if $c_{ij}^{out} = 1$, node j is the major importer for node i, and consequently, the shortage of the export also affects node i itself. Then nodes i and j are too interdependent, and they cannot use the flow w_{ij} to influence each other (contrary to Model 1). This situation can be seen in real trade networks. For instance, some countries of the European Union are too interdependent on each other in terms of traded goods and flows of money between them, so they cannot use the flows to influence each other. This idea supports recent studies which prove that strong trade relations reduce the probability of conflict among the states (Martin et al. 2005, Tanious 2019).

Our Model 2 takes this feature into account. The main idea is that before considering how nodes influence each other, we aggregate matrices of pairwise influence C^{in} and C^{out} into a single interdependence matrix C. Then we calculate the level of interdependence c_{ij} as

$$c_{ij} = c_{ij}^{in} - c_{ij}^{out}. \qquad (1.26)$$

According to Equation (1.26), negative values of c_{ij} show that node i is more dependent on node j than vice versa. Similarly, $c_{ij} > 0$ means that node i influences node j, while $c_{ij} \approx 0$ means that nodes do not influence each other.

Matrix C can be further transformed into matrix $\hat{C}$, which takes into account indirect connections among nodes. Analogous to the first model, this step can be performed by considering different paths or random walks in a graph based on matrix C (see Model 1).

As has been discussed before for Model 1, matrix $\hat{C}$ can be converted into a single index $\hat{c}_i$ which corresponds to the overall level of influence in a network.

3) Model 3: nodes interdependence as a search for influential paths.

In Model 3 we combine the ideas of two previous ones. Model 1 does not take into account that nodes, if they are interdependent, actually do not influence each other. Model 2 does not distinguish the cases when nodes are highly interdependent ($c_{ij}^{in} = c_{ij}^{out} = 1$) and when they are not connected ($c_{ij}^{in} = c_{ij}^{out} = 0$). However, these cases are completely different: in the first case, node i influences node j, and consequently, its neighbors, however, it may not be reasonable to do it, as node i also suffers some losses. In the second case, node i has no power on node j and its adjacent nodes.

In the third model, we attempt to consider all these specific features. The model is based on the following concept: in order to understand whether node i influences node j or not, we consider all possible paths between nodes i and j, and choose a path, P_{i-j}, which maximizes the influence of node i, and, at the same time, minimizes its losses through this path.

Formally, we define the interdependence between nodes i and j as

$$\tilde{c}_{ij} = \max_{P_{i-j}} \left(f^{in} \left(P_{i-j} \right) - f^{out} \left(P_{i-j} \right) \right). \qquad (1.27)$$

where $f^{in}(P_{i-j})$ and $f^{out}(P_{i-j})$ are path strengths which can be evaluated using Equations (1.22) or (1.23). One should note that $f^{in}(P_{i-j})$ and $f^{out}(P_{i-j})$ show how node i is critical and dependent for node j through some path, P_{i-j}.

The idea of Model 3 is also illustrated in Figure 1.8. In Model 2, $c_{AB} = 0$ we use Equation (1.26) to aggregate networks. Since there are no connections from node A to node D, we can derive that $\hat{c}_{AD} = 0$. However, according to this model, $\hat{c}_{AD} = 1-0.1 = 0.9$, there is only one path among nodes A and D in the layer of influence (via node C). Consequently, $f^{in}(P_{A-D}) = 1$ and $f^{out}(P_{A-D}) = 0.1$. Similarly, there are two paths between nodes A and C (direct or via node B), hence, $\hat{c}_{AC} = \max(-0.1, -0.4) = -0,1$, i.e., node A is slightly dependent on node C. Also, Model 1 does not capture this feature as it maximizes the influence with no respect to the other layer.

Similar to previous models, matrix $\hat{C}$ can be converted into a single index $\hat{c}_i$ that corresponds to the overall level of influence in a network.

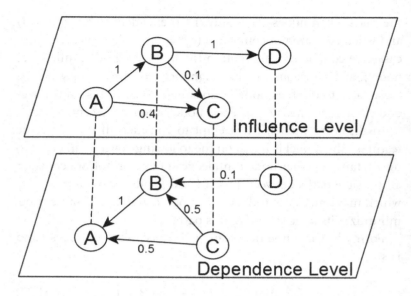

FIGURE 1.8   Illustration of Model 3.

1.4 IMPACT OF INDIRECT CONNECTIONS IN NETWORK STRUCTURES

Over the past decades, the problem of detecting the most important elements in networks has attracted many researchers. Most of the work in this field considers the importance of nodes in networks. One of the most popular ways to indicate pivotal elements in networks is to use centrality measures. However, few studies estimate the most important relations between nodes. In the same way as for the nodes, discovering strong connections can help us to better understand the organization of a network.

There are several methods to evaluate edge centrality. One of the most well-known measures is edge-betweenness centrality (Girvan & Newman 2002). Similar to node centrality, this measure calculates the number of shortest paths passing through each edge. The higher the value of this centrality, the more important the edge is. This measure can be useful in different types of transport networks (traffic, airports, etc.).

Another model of edge importance evaluation is given in De Meo et al. (2012). It is called k-path edge centrality. To evaluate the importance of edges, the model of random walks of length up to k is used. This idea can also be applicable to the information propagation problem.

In Page & Perry (1994), the idea of reliability polynomials is used to evaluate the importance of edges. Edge e_1 is considered to be better in some sense than edge e_2 if the reliability function of the graph without e_1 is less than the reliability function of the graph without e_2.

In Nicholson et al. (2016), the notion of flow is considered in several ways. Authors propose the idea of edge importance based on max flow count, min cut set count, edge flow centrality, flow capacity rate, and damage impact.

Spanning edge centrality is proposed by Mavroforakis et al. (2015). This measure estimates edge importance as the fraction of spanning trees that contain the edge. Classic centrality measures for nodes can also be useful in this problem. For instance, such evaluation can be performed if an initial graph is represented as a line graph, i.e., a graph that represents relations between links (Ortiz-Gaona et al. 2016). However, the complexity of this approach is very high.

We focus on the idea that indirect connections between nodes might play a significant role in network analysis. Our goal is to reveal implicit channels that increase the influence of nodes on each other, as compared to direct intensities of the influence between nodes.

We illustrate the idea of the measure on a graph with three nodes only (see Figure 1.9).

Let us assume that the weights on this graph indicate the intensities of the direct influence between nodes. For instance, this graph can be constructed based on the idea of SRIC or LRIC indices, as described in Section 1.2. Our goal is to detect an edge (or a set of edges) that are implicitly important in terms of the total influence of nodes on each other. We address Section 1.2 in order

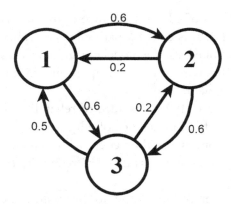

FIGURE 1.9 Example of a graph with intensities of the influence on edges.

to define the influence through simple paths between nodes. One of the possible ways to evaluate the indirect influence through a path is to calculate the product of direct influences on this path (see Equation (1.22)), which can be considered as a joint probability of the influences. The longer a path is, the less contribution it has in terms of the influence.

For the graph in Figure 1.9 one can notice that the direct influence of node 2 on node 1 is less than the indirect influence through edges (2,3) and (3,1), since $0.2 < 0.6 \times 0.5$. The same situation is seen for a connection between nodes 3 and 2: the influence through edge (3,2) is less than through edges (3,1) and (1,2) (see Figure 1.10).

We can see that edge (3,1) appears twice on paths with the indirect influence being higher than the direct influence. Hence, this means that this edge is highly important, as it provides the higher influence for a pair of nodes. If the edge (3,1) is excluded from the graph, the impact for two pairs of nodes declines considerably. Note that edge-betweenness centrality gives equal scores to all edges in this graph. The cut set approach also does not point out the edge (3,1) as the most important one.

Now the model of edge influence estimation will be given more formally. The main idea is to point out edges that increase

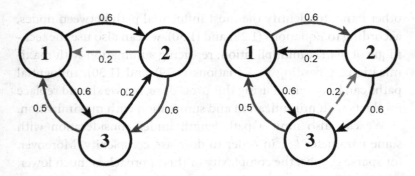

FIGURE 1.10 Indirect influence between nodes 2 and 1 (left) and between nodes 3 and 2 (right).

the influence of nodes on each other, compared to the influence through direct connections between them.

Note by $I(P^t_{i-j})$ the influence of t-path, which can be defined similar to Equation (1.22) as:

$$I\left(P^t_{i-j}\right) = f_{mult}\left(P^t_{i-j}(s)\right) = c_{ik_1(t)} \times c_{k_1(t)k_2(t)} \times \ldots \times c_{k_{s-1}(t)j} \quad (1.28)$$

or similar to Equation (1.23) as:

$$I\left(P^t_{i-j}\right) = f_{min}\left(P^t_{i-j}(s)\right) = \min\left\{c_{ik_1(t)}, c_{k_1(t)k_2(t)}, \ldots, c_{k_{s-1}(t)j}\right\}. \quad (1.29)$$

Now, the problem can be formulated as follows,

$$k^*_{ij} = \arg\max_t I\left(P^t_{i-j}\right),$$

$$w.r.t.\ I\left(P^t_{i-j}\right) > w_{ij}. \quad (1.30)$$

This problem can be solved similarly to the powering adjacency matrix W, as W^l contains the number of paths of length l between pairs of nodes. Generally, when we multiply matrix A by matrix B, we multiply elements of corresponding lines in A to corresponding columns in B sequentially and sum up the products. On the other hand, to identify the most influential path between nodes, according to Equations (1.28) and (1.30), we can also use the technique of matrix multiplication, replacing summation with maximization. According to Equations (1.29) and (1.30), influential paths can be obtained using this procedure, but we should replace products with minimization and summation with maximization.

We can also limit a path length under consideration with some parameter l_{max} in order to decrease complexity. Moreover, for sparse graphs, the complexity of this approach is much lower, which can be observed often in many real networks.

Note the set of solutions of the Equation (1.30) by $P^* = [P_{i-j}^{k^*}]$. This set contains all paths chosen as influential ones. After we collect the information about the most important paths for all pairs of nodes, we reward edges with respect to the number of times they appear in P^*.

Some of the ways to score important edges are described in the algorithm in Table 1.11. For instance, in the algorithm variable, *Score1* contains the number of times each edge is met on the paths from P^*. This measure shows how many times the most influential channels between nodes pass through each edge. Another variable, *ScoreInf*, contains the sum of influence values of the paths to which each edge belongs. It helps us to understand how each edge contributes to the influence calculation within a whole graph. The variable *ScoreContr* is calculated as a contribution of a path that contains edge e, compared to direct influence between considered nodes. *ScoreDiff* takes into account the first two sets of the most influential paths P^* and P^{**} and adds up the difference between influences on these paths. This measure expresses the possible loss in influence if the most influential path is terminated. All these measures can be normalized to corresponding edge attributes (weights or some other external parameters) on the final stage, as well as during the calculation of scores. Other ways to score edges from P^* can be used as well.

As a result, we have described how to detect the most influential paths and proposed several indirect influence assessment

TABLE 1.11   Algorithms for Scoring Edges in Graph G

Algorithm EdgeScoring

INPUT: G, P^*, P^{**} # G – graph
        #P^* – the set of the most influential paths
        #P^{**} – the set of the second most influential paths
for e **in** $edges(G)$: #initialization
    $Score1[e] = 0$
    $ScoreInf[e] = 0$
    $ScoreContr[e] = 0$
    $ScoreDiff[e] = 0$
for i, j **in** $nodes(G)$ # for each pair of nodes
    for e **in** P_{ij}^*: # for each edge on selected paths between i and j
        $Score1[e] += 1$ #add plus 1 score
        $ScoreInf[e] += I\left(P_{ij}^*\right)$ #add the influence of a current path
        $ScoreContr[e] += I\left(P_{ij}^*\right) - weight\,(i, j)$ #add contribution of a current path
        $ScoreDiff[e] += I\left(P_{ij}^*\right) - I\left(P_{ij}^{**}\right)$ #add the difference of the most influential
path
            #and the second most influential path
return $Score1, ScoreInf, ScoreContr, ScoreDiff$

---

measures. All these measures give adequate results for graphs of special forms. For instance, in case of a fully connected graph with equal weights, or a directed star graph, no edge will be scored, as all edges provide the maximal direct influence, and no additional information can be obtained. Our approach can be used to detect edges connecting different components in a graph (bridges). Note that these measures can be applied to networks with many components without component iteration.

1.5 CONCLUSION

We have proposed SRIC and LRIC indices for power evaluation in networks where nodes are heterogeneous and may have additional attributes. These methods are based on the analysis of all simple paths between any pair of nodes.

Contrary to many centrality measures in networks, we consider the individual attributes of nodes. This information helps us identify the most important elements more precisely. These attributes may be obtained externally, or from the network structure and the position of nodes in it.

Another key component of the proposed models is the group influence. In many cases, one node does not have sufficient power to influence another node. However, if nodes are grouped together, they have more power to influence other nodes. Then, the intensity of a pairwise influence should be proportional to the contribution of a node to an influential group.

We have considered long-distance connections between nodes as well. One of the examples is alternative routes chosen in traffic jams (which may be longer by distance but more time-saving). Based on this intuition, we assume that elements may influence each other indirectly, and include many intermediate points. Here we also introduce a parameter of the number of intermediate elements to reduce the complexity of the method.

Additionally, we have extended the LRIC index for the evaluation of influence in a network where a flow in the network may result in nodes becoming too interdependent on each other, and, consequently, have some power against each other using the same flow. The proposed measure is called the interdependence index.

Finally, we proposed several measures of the edge importance assessment. Our goal was to reveal edges that provide higher influence of nodes on each other, compared to direct influence between them. Our approach, which is based on LRIC measures, helps to detect edges that are important to increase the total influence in a network. Elimination of such edges leads to the decrease of the influence between many pairs of nodes.

All presented models are implemented in the SLRIC library, which can be downloaded from the site https://github.com/SergSHV/slric.

CHAPTER 2

Applications

I N THIS CHAPTER, WE present an empirical application of the proposed indices. As SRIC, LRIC, and interdependence indices take into account different features such as individual parameters of nodes, group influence of nodes to a node, and long-range interactions, their application allows us to consider specific aspects of the problem under consideration, thus, detecting hidden influential participants which cannot be elucidated by other measures. We provide several examples of the indices' application to the networks of the foreign claims market, international migration, international trade relations, the global food network, the global arms transfer network, the network of terrorist groups, and the network of international economic journals. We also analyze the correlation among some classical and new indices in these examples using Goodman and Kruskal's γ rank coefficient and the Kendall τ-coefficient. Most of these studies aim at examining the influence of countries in the corresponding networks. However, the list presented is not exhaustive, so the proposed classes of indices can be used to analyze other problems in the network analysis.

DOI: 10.1201/9781003203421-2

2.1 KEY BORROWER DETECTION IN THE GLOBAL FINANCIAL NETWORK

The detection of key borrowers in the foreign claims market has received particular attention in the context of the risk allocation problem. The financial system of a country contains many financial institutions, some of them very important for the system. A failure, or a near failure, of such institutions may rapidly disrupt the whole system and lead to an economic crisis. Thus, the study of connections between institutions, and the identification of organizations, particularly banking systems of systemic importance, present important problems for evaluating financial stability, and support high-quality macroeconomic supervision.

The identification of systemically important elements is inescapable in the context of the risk allocation problem. These studies are generally based on quantitative statistical analysis of indicators for each component of the system, or on the analysis of the sustainability of the network. One common solution is the indicator-based approach. It uses the system of financial indicators estimated by the regulator and measures the bank's contribution to certain activities. For instance, the size of the institution (or the share in the total assets of the system) is considered as a very important parameter from the systemic point of view. Other such parameters to measure the systemic importance of a financial institution might be interconnectedness and substitutability (BCBS 2013, Chan-Lau 2010, IMF 2009, IMF 2010).

An alternative approach to estimate the level of interconnectedness is the use of network analysis. In this case, financial relations can be represented as the system of nodes (financial institutions) and links (e.g., flows of capital) among them. There is a broad range of studies that uses network analysis in the context of stock ownership networks (Garlaschelli et al. 2005), interbank market and payment systems (Furfine 2003), (Iori et al. 2008), financial firms (Chan-Lau 2018), and other financial systems (Battiston et al. 2010). There are also other studies focused on the interconnectedness of the financial system at the national and international level (Allen et al. 2000, Allen et al. 2009). The evolution of the global financial system with respect to its multilayer structure is examined in Korniyenko et al. (2018).

Our methodology has been applied to the network of foreign claims where nodes correspond to the banking systems of countries, and edges indicate the amount of flow. The international financial linkages of the banking systems are collected from the Bank of International Settlements (BIS) (CBS 2020). Our work is based on consolidated banking statistics (CBS) which provide information about foreign claims (e.g., loans, deposits, debt securities, derivatives, etc.) on an ultimate risk basis. According to CBS, claims on an ultimate risk basis are allocated to the country and sector of the entity that guarantees the claims (BIS 2019a). Alternatively, if a bank from country A extends a loan to a company from country B, and the loan is guaranteed by a bank from country C, this loan would be reported as a claim of the country C. CBS are widely used on an ultimate risk basis to gauge the reporting banks' exposures to different countries and sectors (BIS 2019b).

Additionally, the CBS provide data about foreign claims at the regional level, on international organizations as well as unallocated claims. We exclude such information from further analysis since we are focused on cross-country relationships.

The BIS statistics provide quarterly data from the first quarter of 2005 to the first quarter of 2020. The number of reporting countries has increased, from 16 in 2005-Q1, to 23 in 2020-Q1, and includes the G10 countries (Belgium, Canada, France, Germany, Italy, Japan, the Netherlands, Sweden, Switzerland, United Kingdom, and the United States) plus Australia, Austria, Chile, Finland, Greece, India, Ireland, Portugal, South Korea, Spain, and Turkey. Unfortunately, the dataset does not take Chinese banks into account, which are important providers of international bank credit since they do not report these credits to the CBS. According to the BIS Statistical Bulletin, the dataset covers about 93% of total foreign claims and other potential exposures on an ultimate risk basis. The network representation of the largest borrowers (14 nodes, 127 edges) for the end of 2015-Q1 is presented in Figure 2.1.

Key borrowers can be found using classical centrality indices. However, these measures do not take into account two important aspects of the problem: attributes of nodes and the possibility of their group influence. First, nodes in a network that correspond to various banking systems are not homogeneous. For instance, the financial network includes G10 countries, however, it also contains some developing (e.g., Brazil) or less developed countries (e.g., Uganda). Since the same flow amount can be critical for a least developed country, but non-disruptive for a developed country, the individual attributes of nodes should be taken into account. Second, countries can be engaged in different international alliances; they can influence other countries as a group.

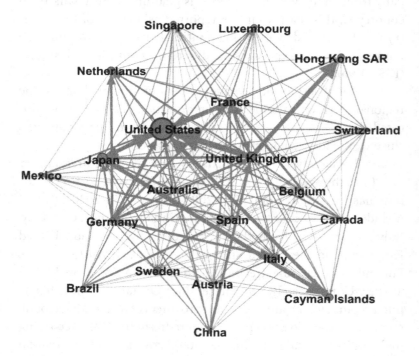

FIGURE 2.1 The largest borrowers in 2015-Q1.

Thus, there is a need to consider a group influence of nodes. For this reason, we applied the SRIC and LRIC indices, but we evaluate classic centrality measures as well, and compare them to SRIC and LRIC measures.

Taking into account the parameters of nodes in the SRIC and LRIC indices needs to define the threshold of influence for each node. One possible way to define it, is to follow the recommendations of the Basel Committee on large exposure limits (25% of Tier 1 capital) (BIS 2018). At the international level, when we deal with banking systems' borrowings, choosing an appropriate threshold level (critical loan amount) is not an easy exercise. We decided not to use loans, but a ratio of the gross domestic product (GDP) of the lending country as a threshold, in order to take the relative size of the loan into account. More precisely, the threshold of influence for each node is defined as 10% of its nominal GDP.

In Table 2.1, we provide a list of countries that were in the top five in 2015-Q1 using at least one of the centrality measures which are the most relevant to the area. Weighted in-degree centrality identifies the main borrowers in the network in terms of the total amount of loans. The PageRank algorithm is based on random walks in a graph, and it assigns high values to nodes that have a higher probability of being visited. An intuitive interpretation of PageRank centrality for financial networks is that the most visited nodes correspond to financial institutions which are highly engaged in financial activities; hence, they may increase the spread of the shock through the system. Eigenvector assigns relative scores to all nodes in the network, based on the concept that connections to high-scoring nodes contribute more to the score of the node in question than equal connections to low-scoring nodes.

Finally, we calculated three versions of the LRIC measure. LRIC (Max) and LRIC (MaxMin) identify the most influential path between nodes and evaluate it in terms of joint 'probability' and flow capacity. LRIC (Sum) considers all possible indirect paths of influence between nodes. One should also note that the

TABLE 2.1 Ranking of Pivotal Borrowers (Top Five) for 2015-Q1 by Centrality Measures

Name	Weighted In-Degree	PageRank	Eigenvector	SRIC	LRIC (Max)	LRIC (Sum)	LRIC (MaxMin)
United States	1	1	1	1	1	1	1
United Kingdom	2	2	2	2	3	3	4
Germany	3	3	5	5	6	10	9
France	4	4	4	6	12	12	8
Cayman Islands	5	7	10	17	5	6	5
Japan	6	5	3	3	11	17	11
Hong Kong SAR	8	9	13	4	2	2	2
China	9	8	14	19	4	3	3
Singapore	14	14	16	25	8	5	6

LRIC index provides a more accurate ranking compared to the SRIC index, as it also considers indirect connection among nodes.

The results of the indices application are as follows: first, a list of countries with large and strong economies, such as the US, the UK, and Germany, was detected. SRIC and LRIC measures also include China in the list of the most influential countries. They have developed financial systems which have high levels of trustworthiness and sovereign ratings, and thus attract a large number of investors.

However, in contrast to classical centrality measures, SRIC and LRIC indices increase the influence of Hong Kong SAR, the Cayman Islands, and Singapore. These countries could be good examples of 'too interconnected to fail' economies. Due to their attractive business environments, well-developed infrastructures, human capital, and positive reputations, these countries stimulate investors to place their assets in their financial systems, which makes the countries important borrowers. Their appearance in the top ranking does not look right at first glance, but it is in line with our initial hypothesis that the greatest influence must have not only the highest number of market participants, but also the most interconnected ones. In other words, for these countries, each individual cash flow is not so significant, but their combination can be critical for the stability of the financial system as a whole. For example, in the case of the elimination of a country from the network, we will most likely not see a chain of cascading failures (because the volumes of interaction are not so great to each country) but will lead to the redirection of financial flows to the other countries, which might affect their overall financial stability. If we consider the individual threshold of each country and the group influence, the list of countries obtained will be critical for the stability of the financial system as a whole, and their elimination will most likely lead to a chain of failure in other countries.

We can also compute the correlation among centrality measures and LRIC indices. Since the position in the ranking is a rank variable, in order to assess the consistency of different orderings, other than the traditional Pearson rank correlation, coefficients should be used. In our work, the idea of Kendall metrics (Kendall 1970), which counts the number of pairwise disagreements between two ranking lists, is applied. Also, we used Goodman and Kruskal's γ rank coefficient, which shows the similarity of the orderings of the data when ranked by each of the quantities (Goodman & Kruskal 1954). The results are provided in Tables 2.2 and 2.3.

We can see that, according to our estimations, the ranking of the LRIC index is highly related to the results of the PageRank index. This fact is confirmed by both our correlation coefficients (the Kendall τ, and the Goodman & Kruskal γ-coefficient). It should be noted that the weighted in-degree centrality also gives similar rankings. Overall, we can see that a correlation coefficient between centrality measures is relatively high, which can be explained by the fact that large borrowers should be more important, as they are more engaged in financial relations. Finally, we can observe that all versions of the LRIC measure have a good correspondence with each other.

However, although the correlation coefficients of most centrality measures and LRIC indices are quite high, it was shown in Table 2.1 that, in contrast to LRIC indices, classical centrality measures do not systemically detect the important countries of the second group (e.g., Hong Kong SAR, Cayman Islands, and Singapore), as they do not consider the individual attributes of nodes and their indirect influence.

A detailed explanation of the data, the choice of thresholds, and the results, as well as the descriptions of related literature, are provided in Aleskerov et al. (2016a, 2020a). A dynamic analysis of the global financial network is presented in Shvydun (2020a).

2.2 NETWORK ANALYSIS OF INTERNATIONAL MIGRATION

The international migration process has several important consequences, both for source and destination countries. First, it

TABLE 2.2 Kendall τ-Coefficient

Name	Weighted In-Degree	PageRank	Eigenvector	LRIC (Max)	LRIC (Sum)	LRIC (MaxMin)
Weighted In-Degree	-	0.984	0.963	0.836	0.824	0.864
PageRank		-	0.951	0.939	0.937	0.933
Eigenvector			-	0.897	0.897	0.866
LRIC (Max)				-	0.998	0.988
LRIC (Sum)					-	0.987

TABLE 2.3 Goodman, Kruskal γ-Coefficient

Name	Weighted In-Degree	PageRank	Eigenvector	LRIC (Max)	LRIC (Sum)	LRIC (MaxMin)
Weighted In-Degree	-	0.753	0.721	0.674	0.69	0.687
PageRank		-	0.828	0.789	0.791	0.768
Eigenvector			-	0.719	0.719	0.681
LRIC (Max)				-	0.976	0.93
LRIC (Sum)					-	0.924

weakens the source countries (e.g., economically, in terms of skills, population, etc.), and provokes new flows of migration from them. Second, it influences the economy and society of countries receiving immigrants. It may also provoke a flow of migration of native people from destination countries. Although the effects of migration, and their costs and benefits, have not been fully explored, it is clear that understanding this process and its key members needs careful evaluation in order to contribute to the development of both source and destination countries.

Migration has been studied in various fields of science. A considerable amount of work was proposed in order to explain the causes of migration flows and the consequences of them. The first studies were focused on the movements of people from rural to urban areas (Smith & Garnier 1838) and provided the fundamental understanding of factors influencing migration (Ravenstein 1889). Lately, several models from other areas have been adapted to the study of the migration process. One of these models – a gravity model of migration – plays a significant role in studying migration flows. The underlying hypothesis of this model is that the level of migration between two territories is positively related to their populations, and inversely related to the distance between them (Tinbergen 1962, Zipf 1946). Several works explore the phenomenon of migration from the perspective of motives to migrate. There was an attempt to explain the migration process by push–pull factors (Lee 1966), and the prospect of the economic theory and human capital approach (Bodvarsson et al. 2015, Sjaastad 1962). All these theories apply different levels of analysis of human migration: the macro level (migration between countries and regions) and micro level (individual). However, they miss the fact that the level of migration between any two countries depends not only on factors related to these two countries, but also on migration flows between other countries.

The migration process can be studied from the network perspective as well. International migration can be represented as a network where the nodes correspond to countries and the edges correspond to migration flows. This approach considers flows between any two countries and shows how the changes in one flow may affect the flows between the other – seemingly unrelated – countries. The application of the network approach to international migration was presented in Davis et al. (2013), Fagiolo & Mastrorillo (2012), and Tranos et al. (2015). However, most of these approaches do not take into account the individual features of each country (e.g., population, economic development), the possibility of their group influence, or chain reactions in the process. A study of the process from the network approach is usually limited to the evaluation of classical centrality measures, which do not take into account the above-mentioned features. The application of our methodology aims to consider all these features and detect the most influential elements of the international migration process.

We used annual information on migrant flows between countries provided by the United Nations (UN 2015) as data. Migration flow was defined as the number of persons arriving in a country, or leaving it, in a given time period. The list of respondents in the database comprised 45 countries. These countries provided information about their incoming and outgoing migration flows. Migration flows for countries not included in the list were accumulated by the statistics of the countries presented in the database. The data was collected through different sources: population registers, border statistics, the number of resident permits issued, statistical forms which people fill out when they change their place of residence, and household surveys. There are three ways to define the country of migrants' origin or destination: by their residence, by their citizenship, or by their place of birth. The preference was given to statistics on residence, because more countries applied this criterion in the 2015 version of the database.

Additionally, as countries apply different criteria to determine international migrant levels and their countries of origin, collect data through different sources, and have various purposes for migration policy, there were some cases of inconsistency in our observations. Overall, in 5% of observations, the data was inconsistent: for the same migration flow data from different countries was not the same (8,672 out of 173,435 observations). We performed several techniques of data aggregation to resolve the problems of inconsistency in our observations. In most of these cases, the difference was not significant, therefore the mean value was taken. However, there were 21 observations where, simultaneously, the minimum value was less than 10, and the ratio between the maximal and minimal values provided was more than 1,000. All these cases were studied individually, and at the end were explained by incorrect statistics in the minimal value data of the country. Thus, only maximum values were taken into account. More details about data preparation are provided in Aleskerov et al. (2016b).

The analysis was applied to data on migration flows in 2013. The international migration network for 2013 is shown in Figure 2.2.

The most important countries can be identified using classical centrality measures. However, these measures do not take into account two important aspects of the problem: attributes of nodes, and the possibility of their group influence. First, nodes in a network that correspond to countries are not homogeneous. For instance, the migration network contains countries like China and India with populations of more than 1 billion; however, it also contains small countries like San Marino, with populations of less than 40,000. In other words, the same flow amount can be critical for a small country and non-critical for more populous countries. Second, migration flows from a particular country A can be insignificant for country B, however, the accumulated migration flow to country B from multiple countries may affect it. Thus, there is a need to consider how a particular group of countries may affect it, and which countries play the major role in such influence. These aspects are taken into account by SRIC and LRIC, thus, we apply these measures to the international migration network.

Since both the SRIC and LRIC indices require a threshold definition, we assume that each country is affected by other countries

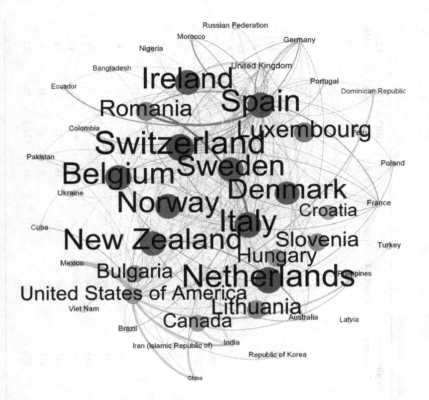

FIGURE 2.2   The international migration network for 2013.

if the number of people immigrating to it exceeds some predefined value. Since all countries differ in terms of population, we decided to use the threshold equal to 0.1% of a population. If this threshold is exceeded, the country is influenced by migrant flows. The aggregation procedure on node-to-node influence was performed with respect to countries' GDP.

In Table 2.4, we provide a list of countries that were in the top ten in 2013, according to at least one of the centrality measures which are the most relevant to the area. Weighted in-degree centrality represents the accumulated incoming flow and thus identifies countries with the largest number of immigrants. Similarly, weighted out-degree centrality identifies countries with the largest

TABLE 2.4 Rankings of Key Players for 2013 by Centrality Measures

Country	Weighted In-Degree	Weighted Out-Degree	Weighted Degree	Weighted Degree Difference	PageRank	Eigenvector	SRIC	LRIC (Max)	LRIC (Sum)	LRIC (MaxMin)
US	1	19	1	1	1	2	22	10	6	10
Italy	2	5	3	4	6	4	11	11	10	16
UK	3	10	4	3	3	1	9	4	9	7
Canada	4	44	5	2	7	12	74	43	37	30
Spain	5	1	2	215	2	3	1	1	1	1
Switzerland	6	12	7	6	5	6	35	54	44	80
Netherlands	7	8	8	10	8	11	17	23	14	27
Sweden	8	21	15	5	11	19	15	38	30	35
Belgium	9	14	10	9	12	9	23	28	19	45
Romania	10	6	6	198	17	5	2	2	2	2
Germany	11	11	9	23	10	7	12	8	4	9
New Zealand	12	16	13	14	4	14	5	15	23	15
France	13	9	12	192	15	8	7	5	3	5
Norway	14	52	23	7	16	23	32	49	45	24
Australia	15	31	22	8	9	20	18	25	21	21
Morocco	18	17	20	166	21	10	8	7	7	6
Poland	23	13	18	210	20	28	4	6	8	4
India	32	2	11	214	26	32	3	3	5	3
Mexico	45	4	14	212	56	40	49	35	40	65
Philippines	53	7	19	211	48	51	16	12	16	11
China	73	3	16	213	55	90	6	9	11	8
Syrian Arab Republic	133	37	49	200	137	137	10	27	28	22

number of emigrants. Weighted degree centrality elucidates nodes with the highest gross migration, while weighted degree difference centrality detects countries with the highest net migration. Note that these classical centrality measures are local, and do not take into account the structure of migration flows. Additionally, we calculated PageRank and eigenvector centralities to consider indirect connections among nodes. An intuitive interpretation of PageRank centrality for migration networks is that the most visited nodes correspond to countries which are highly engaged in the migration process; hence, they play an important role in a network. According to eigenvector centrality, a particular node has a high importance if its adjacent linked nodes have a high importance. In the international migration network, it highlights the countries as 'centers of international immigration', and the countries which are directly linked with them through migration flows. Finally, we applied the SRIC index and different versions of the LRIC measure to evaluate the influence of countries of migration origin on destination countries.

From the results for weighted in-degree centrality, we can conclude that the highest number of immigrants were received by the US, Italy, and the UK. According to the ranking by weighted out-degree, Spain, India, and China, had the highest migrant outflow. Weighted degree ranking highlights the US, Spain, Italy, and the UK, which had the highest gross migration rate. The weighted degree difference, or the highest net migration flow, was found in the US, Canada, the UK, and Italy. Eigenvector and PageRank highlight the 'rich-club' group of countries: the US, Italy, the UK, and Spain. These countries are more involved in the process of migration than others, and in addition, they had flows between them. One should note that all these measures do not consider the populations of countries or how migration flows affect them.

Spain, Romania, India, and Poland had the highest ranking according to the SRIC index. These results are highly similar to the weighted out-degree. As for the LRIC index, Spain, Romania, France, Germany, Poland, and India are at the top of the rankings.

Spain has the highest emigration rate. Romania, India, and France have migration flows to countries with huge populations, and intense migration flows. For instance, France is presented in the rankings of the LRIC index because it has migration flows to Spain (10,548) and to the UK (24,313). Romania also had migration flows to the UK, while there was also a huge flow from India to the US. Poland did not appear among the countries with highest emigration rate (weighted out-degree); however, it had a migration flow of almost 10,000 migrants to Norway, which has a population of around 5 million people. The share of this migrant inflow (0.2%) exceeds 0.1% of the population of Norway. This result is important to consider. Indeed, when a migration flow is more than the level expected by a destination country, it can lead to negative consequences for both migrants and the population of that destination country.

The results introduced by classical centralities and the SRIC and LRIC indices outline the emigration countries. However, the SRIC and LRIC indices additionally introduce the emigration countries which have a considerable share of migrants for the population of the destination country (Poland).

To compare centrality measures, we computed the Kendall τ-coefficient. The highest correlation was observed between PageRank, eigenvector, weighted degree, and weighted out-degree centralities (≈ 0.9). LRIC and SRIC measures have a good correspondence to each other (≈ 0.96), however, their correlation coefficient is equal, on average 0.59, to all classical measures. We can explain this by the fact that these measures also consider the external parameters of nodes (population), contrary to classical measures.

Finally, we computed Goodman and Kruskal's γ rank coefficient among classical centrality measures and the SRIC/LRIC indices. It was observed that SRIC and LRIC indices are highly correlated to the results of the weighted out-degree (≈ 0.84 for SRIC and ≈ 0.77 for LRIC) and weighted degree centralities (≈ 0.83 for SRIC and ≈ 0.8 for LRIC). Eigenvector centrality has a correlation coefficient of about 0.7 with the SRIC and LRIC indices. As for PageRank centrality, it also has a high correlation with the proposed measures ($\approx$0.71 for SRIC and $\approx$0.76 for LRIC).

Overall, our methodology outlined not only the countries with a high number of immigrants or emigrants, but also the countries with migrant outflows which are considerable for the population of the destination country and emigration to the popular destination countries. These results are important in order to provide countries which are highly involved in the process of international migration with a relevant migration policy.

A detailed explanation of the description of related literature, the data, the results, and the dynamic analysis of international migration, is provided in Aleskerov et al. (2016b).

2.3 NETWORK ANALYSIS OF GLOBAL TRADE

International trade is one of the most essential components of the modern economy. It has already been mentioned that exporters and importers may influence each other through trade relations. On the one hand, exporters may decrease the trade volume of products, and consequently influence importers if these products are not manufactured domestically. On the other hand, importers may also influence exporters, as a potential import ban may lead to economic losses for the exporter. Therefore, trade relations make countries increasingly economically interdependent, while trade can be used as a political tool to influence other countries. Thus, it is necessary to identify the most important countries in trade relations.

We can apply our methodology to the real trade network. We used the STAN Bilateral Trade Database by Industry and End-use category (Revision 4) from the OECD (OECD 2020) as the input data. The database presents estimates of bilateral flows between countries from 1990 to 2020. According to the OECD, the latest year shown is subject to the availability of underlying product-based annual trade statistics, thus, we considered the trade network in 2017.

The network is directed and weighted, and contains 198 nodes and 27,264 edges, while the global trade value of goods exported throughout the world amounted to approximately US$19.2 trillion. The network representation of the largest trading partners (41 countries) for 2017 is presented in Figure 2.3.

In Table 2.5 we provide a list of countries that were ranked in the top five in 2017 by at least one of the centrality measures which are the most relevant to the area. Weighted out-degree centrality represents the accumulated outflow, and thus identifies the largest exporters. Note that this measure is local and does not take into account the structure of trade flows. Next, we calculated measures based on the idea of eigenvector centrality: PageRank, eigenvector, and HITS measures. An intuitive interpretation of PageRank centrality for trade networks is that the most visited nodes correspond to countries which are highly engaged in trading relations; hence, they play an important role in a network. The eigenvector centrality measures the importance of a country with respect to the importance of its trading partners. The HITS algorithm

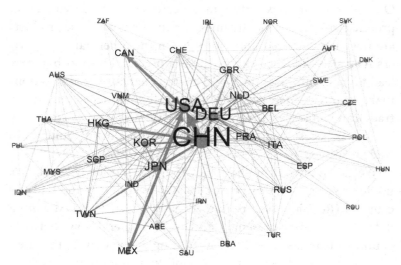

FIGURE 2.3   The global trade network of the largest trading partners for 2017.

TABLE 2.5 Rankings of Countries for 2017 by Classical Centrality Measures

Country	PageRank	Eigenvector	Hubs	Authorities	Weighted Out-Degree
United States (US)	1	17	5	1	2
China (CHN)	2	1	1	3	1
Germany (DEU)	3	16	4	5	3
France (FRA)	4	2	10	10	6
United Kingdom (GBR)	5	3	12	9	10
Japan (JPN)	6	21	6	4	4
Canada (CAN)	7	19	3	7	12
Hong Kong, China (HKG)	11	40	8	2	7
Mexico (MEX)	12	13	2	6	11
Republic of Korea (KOR)	13	10	7	8	5
Czech Republic (CZE)	28	4	33	30	29

assigns two scores – the hub and authority scores – and, according to the model, the node has a high hub score if it links to many authorities. Similarly, a node has a high authority score if it is pointed to by many hubs. Therefore, the hub score is aimed at identifying important exporters, while the authority score detects important importers.

According to Table 2.5, the United States (US), China (CHN), and Germany (DEU) are among the most central elements by at least three indices out of five. France (FRA) and the United Kingdom (GBR) get high scores by PageRank and eigenvector centrality, while Japan (JPN), Canada (CAN), Mexico (MEX), and Hong Kong, China (HKG) are highly ranked by hubs and authorities. If we compare the overall results, the lowest correlation is between the eigenvector centrality and all other measures ($\approx$0.3), and between hubs and authorities ($\approx$0.67), while the correlation of other measures is in the range [0.82–0.92]. However, if we compare rankings, the correlation is high between all centralities: the lowest is between hubs and eigenvector ($\approx$0.78); the highest is between PageRank and authorities ($\approx$0.96).

Although the most important countries can be identified using classical centrality measures, it is necessary to note that these measures do not take into account two important aspects of the problem: the attributes of nodes and the possibility of their group influence. As for previous applications, nodes in a network that correspond to countries are not homogeneous. For instance, the trade network contains countries that have different populations, levels of income, and stages of development, as well as levels of production of some commodities. Thus, it is likely that the same trade flow amount can be significant for one country and noncritical for another. Second, countries can be engaged in different international alliances; they can influence other countries as a group. In other words, there is a need to consider a group influence of nodes. Finally, as is mentioned in Section 1.3, trade relations make countries interdependent, hence, a trade flow between two countries implies that they become too interdependent on each other. These features are taken into account by the proposed measures, therefore we apply the SRIC and LRIC indices as well as three versions of the interdependence measure.

An important aspect of our indices is the choice of the threshold of influence for each node. To define how nodes influence each other, we used the gross domestic product (GDP) of countries. For the SRIC and LRIC indices, we assumed that country i influences country j through its exports if it exceeds a certain percentage of the GDP of country j. For the interdependence measure, we additionally consider the influence of importers on exporters. Thus, we also assume that country i is dependent on the export to country j if it exceeds some level of its own GDP. In other words, we evaluated the threshold of influence (q_i^{out}) or dependence (q_i^{in}) of country i as

$$q_i^{out} = q_i^{in} = \lambda \cdot GDP_i. \qquad (2.1)$$

As for the λ value, we considered different levels $\lambda = 0.1$ (low influence threshold), $\lambda = 0.2$ (medium influence threshold), $\lambda = 0.3$ (high influence threshold). Below we provide the results of our models for a high influence level. In Table 2.6 we provide a list of countries which were ranked in the top five by at least one

TABLE 2.6 Trade Network in 2017: SRIC, LRIC and Interdependence Indices (Top Five, $\lambda = 0.3$)

Country	SRIC	LRIC (Max)	Interdependence Measure		
			Model 1	Model 2	Model 3
China (CHN)	1	1	3	2	2
United States (US)	2	2	1	1	1
Germany (DEU)	3	3	2	3	3
France (FRA)	11	6	4	4	4
United Kingdom (GBR)	21	12	5	7	7
Japan (JPN)	4	4	6	5	5
Italy (ITA)	14	7	11	8	10
Netherlands (NLD)	5	5	7	15	15

centrality measure. Note that we provide only one version of the LRIC index, due to its good correspondence with other versions. According to Table 2.6, China (CHN), the United States (US), and Germany (DEU) are the top three countries with the highest influence levels. China has higher scores than the United States for SRIC and LRIC measures, which can be explained by the fact that China is the largest exporter, and it individually influences 15 other countries, while the United States individually influences only five countries. Moreover, if we calculate the number of countries that China or the US influences in a group with other countries, the total number will be equal to 51 for the US and 55 for China. However, if we assume that importers may also influence exporters through the same trade flow (interdependence of countries), the United States will take first place in the ratings by the interdependence measure.

Overall, the correlation coefficient between all versions of the interdependence centrality is in the range [0.96–1.00] for the scores and [0.8–0.99] for the rankings. The highest correlation is between Models 1 and 3 because Model 2 eliminates many edges. The correlation coefficient between the SRIC and LRIC indices is 0.9. Finally, the LRIC has a higher correlation with the interdependence centrality (≈0.92) compared to the SRIC measures (≈0.79).

We can also note that all models have a good correspondence with classic centrality measures on a trade network. The correlation coefficient with PageRank is from 0.94 to 0.97 for the scores and from 0.87 to 0.95 for the rankings. Other classical measures have lower correlation: ≈0.29 for eigenvector, ≈0.79 for hubs, and ≈0.85 for authorities. However, the proposed models allow us to achieve more accurate results, as they take into account individual attributes of nodes (size, threshold of influence, etc.), a possibility of their group influence as well as nodal interdependence. Moreover, our models provide a more detailed analysis of connections in the network under study. This approach can be used to identify flows, which are important only for a target/source node or for both nodes simultaneously.

More information about this study is provided in Shvydun (2020c).

2.4 INFLUENCE OF COUNTRIES IN THE GLOBAL FOOD NETWORK

Needless to say, food is a critical source of life for all human beings. However, in 2017, 10.9% of the world's population suffered from undernourishment (FAO 2018). A lot of countries are still food insecure, i.e., people in these countries have no physical (social), or economic, access to quality food in sufficient amounts to meet their dietary needs (Simon 2012). There have been several attempts to measure the level of food security (Global Hunger Index (2018), Global Food Security Index (2018), etc.), but most of them are based on statistical factors and do not take into account relations between countries. Unfortunately, there are no general indicators that measure food security and food insecurity, hence, a lot of discrepancy in the assessment of the stability of countries may occur.

We look at this problem from a network perspective in order to reveal the most important participants, as well as the most important relations, in terms of influence. If we look at the problem of food security from a global point of view, and take into consideration the connections between economies, we can observe new factors influencing the food security of each country. However, we do not observe these when we analyze countries separately from each other. Thus, our approach provides an additional insight into the organization of export/import relations.

In this study, we look at the problem of food security from the point of the influence estimation in the food export/import network. We used the World Integrated Trade Solution (WITS) database for the analysis of trade flows between countries (WITS 2018). It is important to determine which products influence food security the most. We addressed this question to experts from the Food and Agriculture Organization (FAO), who recommended that we use three main product categories: wheat, rice, and poultry meat. We matched these categories with a four-digit H1 code system, where code 1001 refers to wheat (wheat and meslin), code 1006 refers to rice, and code 0207 refers to poultry meat. Both wheat and rice are subcategories of the cereals section. Wheat and meslin include durum and other wheat, both as seeds and in other forms. Rice includes rice in the husk (paddy or rough), husked (brown) rice, semi-milled or wholly milled rice, and broken rice. As for poultry, this category includes fresh, chilled, or frozen meat, and the edible offal of fowl, turkeys, ducks, geese, and guineafowl (COMTRADE 2018).

We also analyze the proportion of these three products in the total import of food products of each country. For instance, poultry meat is a major sector for several African countries, such as Benin (on average 25% of the total food import), Tonga (on average 23%), Gabon (on average 15%), etc. As for cereal products (wheat and rice), they play a major role in African and Central Asian territories, for example, imports of wheat to Ethiopia, Tajikistan, Tanzania, Pakistan, etc. represent more than 50% of the total imports to these countries. Similarly, imports of rice to Benin, Guinea, Niger, etc. make up more than 35% of the total imports to these countries. We can see that these products are highly important for the food security of many countries in the world.

Additionally, we have observed exporter and importer reports, and their own versions of flows between them, and there were some cases of inconsistency in the observations. To resolve this issue, we applied a methodology that aims to deal with the problem of mirror statistics (Meshcheryakova 2020).

The most important countries can be identified using classical centrality measures. We applied classical centrality measures from groups of degree centralities and eigenvector centralities to our networks. However, these measures do not take account of an important aspect of the problem – the individual attributes of nodes. Thus, we also calculated the SRIC and LRIC indices.

First of all, we need to define what we understand by the influence in this network. Imagine that country A has been exporting products to country B, but at some point it stopped, or significantly reduced, trading relations (due to political reasons, domestic issues, natural disasters, etc.). The question is to what extent country B suffers from the loss of the import value. The answer definitely depends on the individual parameters of a country, such as production level, consumption, other export/import relations, etc. Since we did not find any detailed information describing countries' production levels, we considered quotas for each country as $q = 15\%$ of the maximal value over export and import.

The results for the network of wheat in 2006 are provided in Table 2.7. We should mention that we present only one version of the LRIC index (Max) and do not provide other versions, due to their high correlation (≈ 0.95) with the LRIC (Max).

According to Table 2.7, North American countries keep the leading positions as exporters of wheat. France and Germany are also strong exporters, by most classical measures. Additionally, former Soviet states and Eastern European countries are quite influential in the network of wheat. Additionally, Australia and Argentina are very important in the network of wheat. The LRIC index also detects these countries as the most influential exporters of wheat.

We should note that the LRIC index can be used to identify the most affected countries. We can see that a lot of African countries are highly dependent on the wheat supply. Some islands, such as the Maldives, Cyprus, and the Netherlands Antilles, are also affected by other countries in terms of wheat imports. Additionally, Finland, Israel, and Portugal belong to the group of dependent countries.

The results for the network of rice in 2006 are provided in Table 2.8.

Unsurprisingly, Thailand is the leading exporter of rice. Into this list we can add India, the United States, China, and other countries. It is interesting that the LRIC index ranks Italy, which has a lot of trading partners in the world, highly. Spain is also one

TABLE 2.7 Classic Centrality, SRIC, and LRIC Measures for the Wheat Network in 2006 (Top Ten)

#	Weighted Out-Degree	Eigenvector	PageRank	Hubs	SRIC	LRIC (Max)
1	USA (4252316)	France (0.771)	Canada (0.185)	US (0.333)	Canada (0.2252)	Canada (0.1626)
2	Canada (3322495)	Canada (0.4122)	US (0.1795)	Canada (0.1758)	US (0.2205)	US (0.1233)
3	France (2867894)	Germany (0.33)	Russia (0.1177)	France (0.133)	France (0.2131)	France (0.1141)
4	Argentina (1686859)	UK (0.1838)	Kazakhstan (0.1164)	Argentina (0.0607)	Australia (0.0604)	Germany (0.0601)
5	Russia (1382376)	Czech Republic (0.1292)	France (0.0375)	Russia (0.0543)	Argentina (0.0571)	Australia (0.0461)
6	Australia (1048044)	Belgium (0.1231)	Germany (0.0265)	Australia (0.0325)	Germany (0.0437)	Ukraine (0.0416)
7	Germany (1044404)	Spain (0.1018)	Ukraine (0.0176)	Germany (0.0316)	Russia (0.0385)	Russia (0.039)
8	Ukraine (619523)	Netherlands (0.0935)	Spain (0.0138)	UK (0.0208)	Czech Republic (0.0296)	Czech Republic (0.0376)
9	Kazakhstan (559467)	Denmark (0.0811)	Australia (0.0124)	Ukraine (0.0204)	UK (0.0246)	Hungary (0.0344)
10	UK (449126)	Ukraine (0.075)	Hungary (0.012)	China (0.0142)	Kazakhstan (0.0246)	Argentina (0.0302)

TABLE 2.8   Classic Centrality, SRIC, and LRIC Measures for the Rice Network in 2006 (Top Ten)

#	Weighted Out-Degree	Eigenvector	PageRank	Hubs	SRIC	LRIC (Max)
1	Thailand (2618261)	Thailand (0.872)	Thailand (0.1568)	Thailand (0.3408)	Thailand (0.4281)	Thailand (0.2154)
2	India (1759473)	India (0.3297)	Niger (0.0959)	India (0.1848)	US (0.1274)	India (0.098)
3	Vietnam (1302816)	Italy (0.2398)	India (0.0889)	Vietnam (0.1467)	Italy (0.1258)	USA (0.0833)
4	US (1279818)	Pakistan (0.1295)	China (0.0752)	Pakistan (0.0979)	China (0.0597)	Italy (0.0759)
5	Pakistan (1165150)	Spain (0.115)	Pakistan (0.0734)	US (0.0769)	India (0.0541)	Pakistan (0.0703)
6	China (426631)	Belgium (0.0938)	US (0.056)	UAE (0.0299)	Netherlands (0.0272)	China (0.0555)
7	Italy (425452)	US (0.0929)	Vietnam (0.0474)	China (0.0297)	Pakistan (0.0265)	Spain (0.0533)
8	Egypt (315621)	China (0.0763)	Japan (0.0374)	Uruguay (0.0145)	Vietnam (0.0238)	Belgium (0.0492)
9	Uruguay (229761)	Netherlands (0.0697)	Italy (0.0249)	Italy (0.012)	Belgium (0.0233)	Netherlands (0.0386)
10	UAE (173087)	Egypt (0.0688)	Australia (0.0215)	Egypt (0.0078)	Egypt (0.0184)	Vietnam (0.0383)

of the leaders by LRIC centrality but is underestimated by other indices.

As for the most affected countries, we can mention some African countries: Zimbabwe, Gambia, Malawi, and Zambia. Moreover, some European countries, such as Estonia, Latvia, Poland, and Ireland, are also highly dependent on the import of rice. The results for the network of meat in 2006 are provided in Table 2.9.

According to Table 2.9, the largest exporter of poultry meat, and the most important participant, is Brazil. The US is detected as one of the key exporters by most of the centrality measures, while the LRIC values it very close to Poland, in fifth and sixth places respectively. The LRIC index also reveals Belgium and the Netherlands as the most influential countries in the network of poultry meat.

We also calculated the Kendall correlation between all considered centrality measures. We provide the results for a network of wheat in order to understand which measures give close ranking, and which measures rank countries differently (see Table 2.10).

We can see that some of the measures are well correlated. For instance, hubs correlates with out-degree measures at the rate of approximately 0.92. The group of eigenvector centralities correlates at the rate 0.76–0.79 which implies that it is important to look at the results of each measure from this group separately. As for the SRIC and the LRIC, these measures are correlated with each other at the rate of approximately 0.8.

Overall, social network analysis allows us to detect the most influential and the most dependent countries in terms of trade. We apply classical centrality measures that help us to reveal central elements. As expected, the indices elucidated developed countries with high production levels and stable agro-food systems. The main disadvantages of classical measures are that they do not consider individual attributes of nodes, which are very important for networks where the actors are countries. Thus, we applied the

TABLE 2.9 Classic Centrality, SRIC, and LRIC Measures for the Meat Network in 2006 (Top Ten)

#	Weighted Out-Degree	Eigenvector	PageRank	Hubs	SRIC	LRIC (Max)
1	Brazil (3046045.113)	Netherlands (0.5391)	Brazil (0.2872)	Brazil (0.2872)	Brazil (0.2822)	Brazil (0.1806)
2	US (2279705.3545)	Belgium (0.4724)	US (0.2136)	US (0.2136)	US (0.2084)	Belgium (0.0989)
3	Netherlands (1371558.8635)	Brazil (0.4114)	Netherlands (0.1306)	Netherlands (0.1306)	Netherlands (0.1348)	Netherlands (0.0958)
4	France (961876.6605)	France (0.3568)	France (0.0721)	France (0.0721)	Belgium (0.0889)	France (0.0867)
5	Germany (649710.374)	Germany (0.278)	Belgium (0.0484)	Belgium (0.0484)	France (0.0731)	US (0.0768)
6	Belgium (609154.0445)	Poland (0.2538)	Poland (0.0448)	Poland (0.0448)	Poland (0.0454)	Poland (0.0767)
7	Poland (527008.2225)	Hungary (0.1423)	Germany (0.0418)	Germany (0.0418)	Germany (0.0337)	Germany (0.0614)
8	Hungary (311753.64)	UK (0.1156)	Hungary (0.0214)	Hungary (0.0214)	Hong Kong (0.0326)	Hungary (0.0541)
9	UK (303050.2395)	Italy (0.0904)	Chile (0.0174)	Chile (0.0174)	Denmark (0.0253)	Italy (0.0292)
10	Italy (272693.3285)	Denmark (0.0309)	China (0.0161)	China (0.0161)	Singapore (0.012)	UK (0.0289)

TABLE 2.10 Kendall Correlation of Centrality Measures for a Network of Wheat in 2006

	Weighted Out-Degree	Eigenvector	PageRank	Hubs	SRIC	LRIC (Max)
Weighted Out-Degree	-	0.763	0.824	0.917	0.710	0.727
Eigenvector		-	0.776	0.762	0.610	0.715
PageRank			-	0.784	0.727	0.714
Hubs				-	0.664	0.715
SRIC					-	0.803

SRIC and LRIC indices that allowed us to detect several groups of countries with direct, as well as indirect, influence on the routes of different levels in the food network.

2.5 INFLUENCE OF COUNTRIES IN THE GLOBAL ARMS TRANSFERS NETWORK

The transfer of weapons and other military equipment plays a central role in security studies. The results of these studies might be a really pressing matter for policymakers and political activists. It was justified that arms are employed as a potential instrument of influence (Sislin 1954). Arms transfers are also significant and can be used as positive predictors of the increased probability of war (Craft & Smaldone 2002). The arms trade should be understood from both economic and political points of view. However, network-analytic approaches to arms transfers are surprisingly rare. Network analysis in this field was used for primarily descriptive purposes, especially for visualization (Kinsella 2003) and calculating basic statistics, and further used in standard econometric models (Akerman 2014). Very few works consider the arms trade network as a dynamic process (Thurner et al. 2019) or take into account how countries interact with each other with respect to different armament categories. In this regard, we applied centrality indices to understand how countries influence each other globally, in different armament categories, and which countries are the most influential players.

To analyze the global arms trade network, we use the SIPRI Arms Transfers Database from the Stockholm International Peace Research Institute (SIPRI) (SIPRI 2019). The database contains information on all transfers of major conventional weapons from 1950.

Each arms deal is characterized by the type and number of weapon systems ordered and delivered, the years of those deliveries, and their financial values. SIPRI has developed a unique pricing system to measure arms transfers using a common unit – the SIPRI trend-indicator value (TIV). According to Simmel et al. (2012), the TIV measures transfers of military capability rather than the financial value of arms transfers. Additionally, all weapons and other military equipment are combined in the following armament categories: air-defense systems, aircraft, armored vehicles, artillery, engines, missiles, naval weapons, satellites, sensors, ships, and other weapons.

We constructed the global arms transfer network for each year where nodes correspond to countries, and edges indicate the accumulated TIV measure of arms transfers. In this study we decided to consider each armament category individually, otherwise the interpretation of indirect paths would be highly controversial. In other words, a sharp decline in trade of one armament category (e.g., aircraft) from country A to country B will not necessarily lead to a decline in the trade of another armament category (e.g., ships) from country B to country C. However, it is likely that the decline in trade of one armament category (e.g., aircraft) from country A to country B will lead to a decline in trade of the same armament category from country B to country C due to the deficit in country B.

To define the most important elements in the global arms transfers network, one can apply various centrality measures. For instance, weighted out-degree centrality, which is a local measure, identifies the largest exporters of a particular commodity. Similarly, PageRank identifies the most visited nodes that correspond to countries which are highly engaged in the arms trade. We calculated these measures as well as other centrality measures, however, since in this work we are focused on the influence in a network, the results are provided for an LRIC measure. The choice of LRIC measure can be explained by the following:

- Nodes in a network are countries, which are not homogeneous: the same flow amount can be critical for one country and not significant for another;

- Country A with high export values can trade weapons to fewer countries, compared to country B;

- Countries are engaged in various military alliances, thus, they may influence other countries as a group;

- Very long connections do not play an important role in annual data, as the contagion effect requires time in order to affect distant members.

An important aspect of the LRIC index is the choice of the threshold of influence for each node. In general, trade relations among countries cannot be considered separately from the information of the level of production in these countries. For instance, country A may have high import values, and consequently, be highly dependent on other countries according to the network structure. However, if the total import accounts for a small proportion of the overall production level, none of exporting countries actually influences country A. On the other hand, if the level of production in country A is very low, the country becomes highly dependent on its exporters. Thus, the production level could be a good proxy for the threshold of influence.

Unfortunately, we did not find any detailed information describing countries' production levels of a particular armament category. Therefore, we take this feature into account, and calculate the threshold of influence of country i as

$$q_i^\alpha = \lambda \cdot \max\left(\sum w_{ki}^\alpha, \sum w_{ik}^\alpha \right) \tag{2.2}$$

where w_{ki}^α (w_{ik}^α) are the TIV measures of all weapons from armament category α traded from country k to country i (country i to country k), and λ is an additional parameter, $0 < \lambda \leq 1$. The main idea of the equation is that if the total export of country i is higher than its total import, country A will be less dependent on imports, as it sells more than it buys. As for the λ value, we consider $\lambda = 0.25$ (low influence threshold), $\lambda = 0.5$ (medium influence threshold), $\lambda = 0.75$ (high influence threshold).

Thus, we evaluated the influence of countries for each armament category with respect to parameter λ. Since all threshold values provide similar results, we present the results on the most important countries for the two largest armament categories – aircraft and ships – with respect to the medium influence threshold in Figures 2.4 and 2.5. Note that since all versions of the LRIC measure showed a good correspondence to each other, we considered the LRIC (Max) index.

Overall, the United States, the Soviet Union, Russia, and the United Kingdom, which are traditionally the largest exporters of arms transfers, are among the most influential members in most armament categories. However, these countries are not necessarily the most influential participants across all categories of armaments: for instance, Czechoslovakia is among the leaders in the aircraft trade in the late 1950s (see Figure 2.4), while Germany has been one of the most influential members in the ship trade

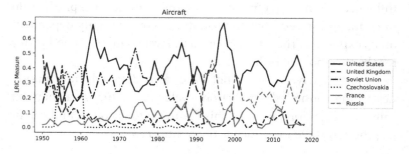

FIGURE 2.4   Influence of countries in the aircraft trade network.

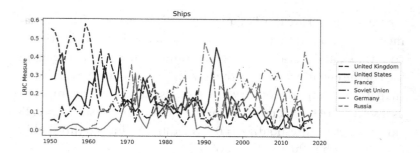

FIGURE 2.5   Influence of countries in the ship trade network.

network from 1970 (see Figure 2.5). We have also examined other armament categories and identified that Switzerland was the most influential country between the 1960s and the 1980s in the sensors trade network, Germany was the most influential country in the late 2000s in armored vehicles.

We can observe two main features. First, some countries specialize in particular armament categories and play an important role in arms transfers. Second, although the United States, the Soviet Union, and the United Kingdom monopolized some categories in terms of LRIC values between 1950 and the 1970s, their relative importance has decreased, and now it is comparable to some other exporters.

Overall, although there are some evident leaders such as the United States, the United Kingdom, the Soviet Union, and Russia, the LRIC measure also detected some other countries that have high influence values for some armament categories.

More information about related works, the data, and the results, are provided in Shvydun (2020b).

2.6 POWER DISTRIBUTION IN THE NETWORKS OF TERRORIST GROUPS

We have discussed the application of the SRIC and LRIC indices to the analysis of interactions among countries. However, the proposed measure can also be applied to an analysis of other entities, for instance, terrorist groups.

Since 9/11, terrorism has become a global issue for the 21st century. Terrorist organizations have gained high influence in political processes and decision-making. Terrorist groups tend to cooperate with other groups – they can form alliances to perform significant operations. This gives a background to construct and analyze a global network of terrorist organizations. Obviously, the structure of this kind of network is not homogenous – some groups have more allies than others, some groups are more powerful and influential. Moreover, terrorist organizations try to influence other groups and gain power over them to avoid competition.

These considerations bring us to the study of power and influence of terrorist organizations. Terrorist networks are widely used in the analysis of international terrorism. However, the vast majority of the existing literature focuses on the ties between individuals within a terrorist network, rather than on the connections between the groups of terrorists (see, e.g., Bacon 2018).

The coalitions between terrorist organizations have been studied in Karmon (2005), pointing out that coalitions usually consist of two groups; larger coalitions are possible if groups are geographically close. Another research related to terrorist intergroup cooperation (Asal et al. 2015) studied ties between organizations and their characteristics. To find the connection between lethality and terrorist alliances, Horowitz & Potter (2014) studied the intergroup cooperation of terrorists. They claim that terrorist groups are used to form alliances instead of acting alone.

Pedahzur & Perliger (2016) studied terrorist groups' networks from another point of view. They showed that a network structure influences groups' effectiveness, which is associated with a large number of hubs and subgroups in a network. In Perliger (2014) a dataset of 18 terrorist networks was used to determine the factors that affect network productivity and durability. Further, Bond (2010) examined the relationship between terrorist groups' (violent non-state actors) power, identity, and intergroup cooperation. It was observed that only 3% of terrorist groups' pairs cooperated with each other more than twice.

Mainas (2012) described basic measures for the investigation of terrorism. These measures applied to terrorist groups' networks are used to define highly active actors or organizers. Medina (2014) applies these measures to the Islamist terrorist network. He concludes that this network is resilient and efficient even if important nodes are removed. Ouellet et al. (2017) calculated Al Qaida's size, density, clustering coefficient, and degree, before and during the War on Terror launched after the September 11 attacks. Latora & Marchiori (2014) proposed a method to identify the critical components of a network. The critical nodes are key players which can be considered as a target to disrupt the terrorist network.

To construct a network of terrorist groups we use the Global Terrorism Database (GTD) which is maintained by the START Consortium (National Consortium for the Study of Terrorism and Responses to Terrorism 2019). The database provides worldwide information on terrorist attacks from 1970. In this study, we used information on terrorist attacks between 2001 and 2018.

In general, there are a total of 1,479 terrorist organizations which perpetrated attacks during the observed period. Using the GTD, we found 1,554 terrorist attacks committed by two or more groups during 2001–2018, considering groups were tied if they acted together. Then we constructed an undirected network with 514 nodes (terrorist groups) and 669 edges (ties between groups). The network is shown in Figure 2.6.

According to Figure 2.6, the global terrorist network includes more than 510 terrorist organizations in different regions of the world. Obviously, they all cannot be connected with each other. For this reason, we analyzed each connected component separately. We have observed that the largest component consists of 304 terrorist groups which accomplished more than 34,600 attacks in total, while the size of other components does not exceed 14 groups. Thus, in this study we present the analysis of the largest connected component, as other components are too small and contain evident leaders.

According to the database, 37 groups of the largest component committed more than 100 terrorist attacks between 2001 and 2018; 69 groups committed more than 20 attacks, while there were 44 groups that organized only one attack. These groups were active in several regions: the Middle East, North Africa, Central Asia, South Asia, Southeast Asia, Sub-Saharan Africa, Europe, and North America. The majority of these groups are religious and separatist; there are also leftist groups – the Communist Party of India (Maoist), the New People's Army, and the United

FIGURE 2.6   Illustration of global terrorist network (2001–2018).

Liberation Front of Assam are the most active and all come from India. Other groups can be considered as ethnic, such as the Kurdistan Workers' Party, Baloch Republican Army, and Al-Aqsa Martyrs Brigade.

As for the highest frequency of the attacks, the Taliban committed 7,336 attacks – more than any other terrorist groups did. The second most active group, the Islamic State of Iraq and the Levant (ISIL), was active only in 2013–2018, and organized 5,034 attacks in 27 countries. Both the Taliban and ISIL are religious terrorist groups. The third group of the subgraph, Boko Haram, is religious and separatist simultaneously. It committed 2,355 terrorist attacks during 2009–2018. The next most active groups are Al-Shabaab (2,336 attacks) and the Communist Party of India (Maoist) (1,895 attacks). The last groups in the subgraph which committed more than 1,000 terrorist attacks are the Maoists

(1,683 attacks), the New People's Army (NPA 1,341 attacks), and Tehrik-i-Taliban Pakistan (TTP, 1,218 attacks). Finally, it is worth mentioning Al-Qaida and its branches in the Arabian Peninsula (AQAP), in Iraq, in the Islamic Maghreb (AQIM), in Yemen, and in the Indian subcontinent, altogether committed 1,551 terrorist attacks.

Analyzing the frequency of terrorist attacks organized by 510 groups of the network (alone or in cooperation), we observe that terrorist activity has been changing over the observation period. We can distinguish at least three periods of the network: 2001–2007 with fewer than 1,000 attacks per year, 2008–2011 with more than 1,000 attacks, and increasing activity in 2012–2018 with more than 2,000 attacks. Hence, we decided to build three different networks, considering terrorist groups that were tied with other groups during these periods, and analyzed each network separately.

We considered unweighted and weighted networks. For unweighted networks, two nodes were connected if they organized at least one attack. For the weighted network, we used the number of joint terrorist attacks and the number of victims as edge weights.

To analyze the influence of terrorist organizations we applied classical centrality measures, such as degree, betweenness, closeness, and eigenvector. Degree centrality identifies organizations with the largest number of partners. Betweenness centrality detects terrorist groups that are located between other nodes and that act as connectors. Closeness centrality elucidates organizations which are located in the center of the network. Eigenvector centrality measures the importance of a node with respect to the importance of its neighbors.

The results of classic measures for the unweighted network are presented in Table 2.11.

Table 2.11 provides a list of terrorist groups that were ranked in the top four by at least one centrality measure. According to the list, the Islamic State of Iraq and the Levant (ISIL) is among the

TABLE 2.11 Classic Centralities (Unweighted Network, Top Four, 2012–2018). The Rank of the Group is Given in Parentheses

Group Name	Centrality (Rank)			
	Eigenvector	Closeness	Betweenness	Degree
Islamic State of Iraq and the Levant (ISIL)	0.41 (1)	0.277 (3)	0.55 (1)	35 (1)
Al-Nusrah Front	0.38 (2)	0.226 (18)	0.05 (21)	21 (3–4)
Ahrar al-Sham	0.31 (3)	0.188 (126)	0.01 (46)	17 (6)
Free Syrian Army	0.29 (4)	0.224 (23)	0.02 (36)	14 (8)
Hamas (Islamic Resistance Movement)	0.08 (18)	0.283 (1)	0.31 (3)	12 (11)
Kurdistan Freedom Hawks (TAK)	0.05 (29)	0.271 (4)	0.18 (7)	3 (88)
Al-Qaida	0.02 (59)	0.281 (2)	0.51 (2)	11 (14)
Lashkar-e-Jhangvi	0.008 (90)	0.258 (5)	0.26 (4)	21 (3–-4)
Tehrik-i-Taliban Pakistan (TTP)	0.007 (91)	0.223 (25)	0.09 (14)	27 (2)

most influential groups, according to all centralities in the largest component for unweighted networks. The Al-Nusrah Front, Hamas (Islamic Resistance Movement), Al-Qaida, and Lashkar-e-Jhangvi are also influential according to two centralities in the largest component.

For weighted networks, we apply the SRIC and LRIC indices as they consider individual attributes of nodes. Since edges in a network correspond to the mutual actions of terrorist groups, it is necessary to understand how important these connections are for each group. For instance, if group A accomplishes 1,000 attacks alone and only 10 attacks with another group, B, it is likely that this connection is not so important for group A. On the other hand, if all terrorist attacks of group B are accomplished with group A, then this connection is significant for group B. Thus, group A may influence group B. This feature is not taken into account by classical centrality measures; hence, we do not provide the result for classical measures and apply the SRIC and LRIC indices.

An important aspect of the SRIC and LRIC indices is the choice of the threshold of influence for each node. In our work, the threshold parameter q_j is defined as the share of the total number of attacks/victims that node j accomplished. We provide the results for three levels of influence: low (5% of all attacks), medium (30% of all attacks), and high (60% of all attacks). Note that since all versions of the LRIC measure showed a good correspondence to each other, we consider only the LRIC (Max) index.

In Table 2.12 we present a list of terrorist groups that were ranked in the top three by at least one centrality measure during 2012–2018.

According to Table 2.12, if the threshold of influence is low, the Taliban and Tehrik-i-Taliban Pakistan (TTP) are the most influential groups. However, the influence of the Taliban decreases with the increase of threshold value, the TTP remains the most influential group for medium levels. Overall, we can observe that the Al-Nusrah Front is among the top three influential groups according to all levels of influence (except for the SRIC, 5%). We

TABLE 2.12 SRIC and LRIC Indices, 2012–2018 (Weighted Network, Attacks, Top Three). The Rank of the Group is Given in Parentheses

Group Name	Centrality (Rank)					
	5%		30%		60%	
	SRIC	LRIC	SRIC	LRIC	SRIC	LRIC
Taliban	0.071 (1)	0.034 (8)	0.042 (7)	0.03 (10)	0.003 (50)	0.002 (47)
Islamic State of Iraq and the Levant (ISIL)	0.07 (2)	0.038 (5)	0.049 (6)	0.032 (8)	0.027 (12)	0.02 (13)
Tehrik-i-Taliban Pakistan (TTP)	0.054 (3)	0.052 (1)	0.053 (3)	0.04 (6)	0.023 (14)	0.016 (16)
Lashkar-e-Jhangvi	0.05 (4)	0.043 (3)	0.019 (18)	0.024 (12)	0.031 (9)	0.026 (11)
Lashkar-e-Taiba (LeT)	0.026 (6)	0.027 (9)	0.052 (4)	0.056 (3)	0.014 (22)	0.004 (40)
Al-Nusrah Front	0.022 (10)	0.045 (2)	0.078 (1)	0.076 (1)	0.097 (1)	0.078 (3)
Hizbul Mujahideen (HM)	0.014 (19)	0.019 (12)	0.055 (2)	0.063 (2)	0.022 (15)	0.013 (19)
Badr Brigades	0.0086 (31)	0.001 (144)	0.025 (11)	0.0226 (14)	0.084 (2)	0.118 (1)
Asa'ib Ahl al-Haqq	0.0085 (32)	0.002 (116)	0.024 (12)	0.0027 (13)	0.077 (3)	0.114 (2)

can also observe that ISIL and the TTP are also in the top 16, according to all thresholds of influence.

Finally, the results of the SRIC and LRIC indices, if the network is constructed according to the total number of victims, are provided in Table 2.13. We present a list of terrorist groups that were ranked in the top three by at least one centrality measure.

Again, the Taliban is one of the most influential groups according to all levels of the threshold, while the Al-Nusrah Front also influences other groups for medium and high levels. The Haqqani Network, the Khorasan Chapter of the Islamic State, Lashkar-e-Taiba (LeT), and Ahrar al-Sham, are also in the top ten according to all levels of influence and may be considered as the most influential groups.

Overall, the SRIC and LRIC allow us to identify groups according to different threshold values. If we apply the LRIC index for the largest component of the network, we can observe a difference between direct and indirect influence. Note that groups that influence the network indirectly do not usually have a strong direct influence on the network. Indirectly influential groups may remain unnoticed by other centrality measures, but they still have access to resources and communication channels to spread information, and an ability to create new connections.

Thus, using the SRIC and the LRIC we can determine influential terrorist groups which could not be recognized by classic centrality indices. In that way, the knowledge about the power of non-state actors and the dynamics of their influence can be expanded. More information about this study is provided in Aleskerov et al. (2020b).

2.7 NETWORK OF INTERNATIONAL ECONOMIC JOURNALS

In this section, we provide an application of centrality measures to evaluate the importance of economic journals. Our main goal is to detect important international economic journals, based on cross-citation networks.

TABLE 2.13 SRIC and LRIC Indices, 2012–2018 (Weighted Network, Attacks, Top Three). The rank of the group is given in parentheses

Group Name	Centrality (Rank)					
	5%		30%		60%	
	SRIC	LRIC	SRIC	LRIC	SRIC	LRIC
Haqqani Network	0.29 (1)	0.064 (4)	0.05 (6)	0.051 (6)	0.06 (4)	0.05 (4)
Khorasan Chapter of the Islamic State	0.11 (2)	0.093 (1)	0.03 (11)	0.026 (12)	0.03 (9)	0.028 (8)
Lashkar-e-Taiba (LeT)	0.1 (3)	0.057 (6)	0.04 (7)	0.052 (5)	0.04 (7)	0.049 (5)
Tehrik-i-Taliban Pakistan (TTP)	0.05 (4)	0.088 (2)	0.12 (2)	0.055 (4)	0.027 (11)	0.017 (16)
Taliban	0.04 (5)	0.08 (3)	0.13 (1)	0.11 (2)	0.18 (1)	0.15 (1)
Al-Nusrah Front	0.03 (9)	0.039 (7)	0.09 (3)	0.12 (1)	0.16 (2)	0.14 (2)
Ahrar al-Sham	0.02 (10)	0.033 (8)	0.08 (4)	0.1 (3)	0.07 (3)	0.08 (3)

One of the main scientometric indicators most often used to rank journals is the impact factor (IF), which determines the value of a publication by calculating the average number of citations of a journal over a certain period of time (usually in the two preceding years). For instance, the international database Web of Science Core Collection (WoS) provides information about the values of two- and five-year IFs.

The research interest in scientometric assessment is supported by practical reasons: journal indicators, along with the Hirsch-index, might, directly or indirectly, determine career growth, wages, and access to grants, for many scientists (Hicks et al. 2015). At the same time, the most popular and demanded indicator – the Thomson Reuters impact factor – has rightly been criticized over a long period, both by experts in scientometrics (Seglen 1997) and by the scientific community (San Francisco Declaration 2013). Thus, there have been proposed alternative methods of journal ranking (Bollen 2005, Leydesdorff & Bornmann 2011). A detailed review of various kinds of scientometric indicators of scientific journals is presented in Waltman (2016).

One of the most popular alternative methods of studying the impact and level of scientific journals is to use a network approach. In this case, the problem can be represented as a cross-citation network where nodes correspond to journals, and edges represent information about the total number of citations of one journal by another. Thus, the citation network is weighted and directed. In Waltman & Yan (2014) the PageRank algorithm was used to evaluate the influence of journals. In Guerrero-Bote & Moya-Anegón (2012) the SCImago journal rank algorithm is used to calculate the Scopus quartiles.

In this work, we apply our models for cross-citation networks to evaluate the importance of international scientific journals in economics, based on the WoS database. According to the database, all journals are assigned to one or more categories (56 categories in total). The total number of journals belonging to the category 'economics' was 333. This includes research from different subcategories including economic theory, econometrics, economic integration and globalization, economics of developed and developing countries, monetary and financial problems, economics of sectoral markets (health care, agriculture, etc.), management and marketing issues, interdisciplinary research, methodology and the methodology of teaching economic disciplines, and economic reviews. Obviously, the level of these journals as well as their degree of influence is different.

The cross-citation network was constructed for 2014. More precisely, we consider citations published in 2014 of papers published between 2011 and 2014. Thus, the publication window is four years, and the citation window is one year. In general, it is possible to count how many times one journal cites another throughout the history of the existence of both periodicals, but we choose a fixed period for both citing journals and cited ones, in order to examine the current situation.

To test our methodology, we selected the first 100 out of 333 economics journals, which, according to Journal Citation Reports (JCR), are the most authoritative, according to the IF. It is necessary to mention that the *Quarterly Journal of Economics*, which has a two-year IF of 6.654 and a five-year IF of 9.794, is the most authoritative journal according to the IF. The last one in our sample is the *Journal of Agricultural Economics* which has a two-year IF of 1.258 and a five-year IF of 1.898. The illustration of the citation network of economic journals is provided in Figure 2.7. The size of nodes corresponds to the total number of citations.

Next, we calculate centrality measures most relevant to the area. Weighted in-degree centrality identifies journals with the highest number of citations. Note that this classical centrality measure is local and does not take into account the structure of citations or the quality of citing journals. Thus, we calculated PageRank and eigenvector centralities that consider indirect connections among nodes. An intuitive interpretation of PageRank centrality for a citation network is that the most visited nodes correspond to journals which are highly engaged in the citation process; hence, they

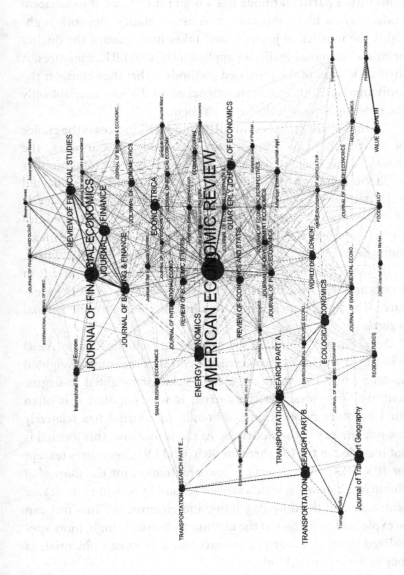

FIGURE 2.7   An illustration of a citation network.

play an important role in a network. According to eigenvector centrality, a particular node has a high importance if its adjacent nodes have a high importance. A cross-citation network highlights the most cited journals and takes into account the quality of citing journals. Finally, we applied SRIC and LRIC measures. A distinct feature of the proposed methods is that they consider the individual attributes of each journal and take into account only the most significant links between them.

An important aspect of the SRIC and LRIC indices is the choice of the threshold of influence for each node. In our work, the threshold parameter q_j is defined as the share of the total number of citations to journal j. In other words, the journal is affected if q_j citations is exceeded. We provide the results for three levels of influence: low (10% of current number of citations), medium (50%), and high (80%). As for group influence, we limit the maximum size of groups to five, as large groups are unlikely to appear.

In Table 2.14 we provide a ranking of international journals in economics that were in the top three by one of the centrality measures. The sign "-" means that the centrality index for the journal is zero.

According to Table 2.14, the *American Economic Review* (AER) is in the first place by two classical centrality measures (weighted in-degree and PageRank). The high value of weighted in-degree centrality indicates that the journal is very popular: it is often cited by other publications, although the journal has relatively few references to other journals. At the same time, this journal is not included in the top three for SRIC and LRIC measures (except for 10% SRIC). We observe a similar situation for the *Journal of Financial Economics*, which ranked second by weighted in-degree centrality and third by PageRank and eigenvector. This fact can be explained as follows: if the citation threshold is high, more specialized groups of journals are attributed as most influential, as they actively cite each other.

As for the eigenvector centrality, the *Journal of Banking & Finance* takes the first place. This journal is connected to many

TABLE 2.14 Ranking of International Journals in Economics for 2014

Journal Name	Weighted In-Degree	PageRank	Eigenvector	SRIC			LRIC (Max)			LRIC (Sum)		
				10%	50%	80%	10%	50%	80%	10%	50%	80%
American Economic Review	1	1	4	1	32	39	9	12	12	25	22	14
Journal of Financial Economics	2	3	3	9	9	46	46	23	40	57	20	40
Journal of Finance	3	4	6	30	43	52	54	28	–	63	30	–
Energy Economics	4	20	2	4	2	5	1	1	15	4	1	15
Transportation Research Part A: Policy and Practice	11	30	18	18	11	12	17	6	2	1	5	1
Journal of Banking & Finance	10	34	1	2	1	10	4	2	4	18	2	6
Ecological Economics	12	26	7	5	3	3	3	3	1	8	4	4
Journal of Economic Behavior & Organization	27	44	9	3	13	60	6	18	33	9	15	34
Transport Policy	25	56	41	52	61	29	38	9	3	12	9	2
Journal of Transport Geography	6	24	12	14	7	2	44	11	7	28	17	3
International Review of Economics & Finance	20	75	8	38	29	19	2	25	45	29	25	45
Econometrica	7	2	10	23	59	40	52	54	29	61	57	29
Pharmacoeconomics	43	71	95	11	4	1	53	17	5	79	11	5

influential journals, consequently, it can seriously impact all journals. It is also in top place according to the SRIC index at $q = 50\%$ and in the top four for almost all versions of SRIC and LRIC measures.

According to the SRIC index, the AER journal is the first for $q = 10\%$, however, for $q = 50\%$ and $q = 80\%$ it did not enter the top three, being respectively in 32nd and 39th place.

The application of the SRIC index showed that highly influential journals by classical centrality indices are not included in the top three. On the contrary, *Pharmacoeconomics, Journal of Transport Geography*, and *Ecological Economics*, which are not included in the top three by classical measures, are included in the top three according to SRIC at $q = 80\%$.

Additionally, we can observe that *Energy Economics*, which is in the top four by weighted in-degree and eigenvector centralities, is in the top five by most versions of the SRIC and LRIC indices. Interestingly, this journal takes the 20th place according to the PageRank centrality.

We have also identified a journal, which is considered as influential only for a low threshold value ($q = 10\%$): the *International Review of Economics & Finance*. This journal does not cite a large number of other journals.

As for the high threshold value ($q = 80\%$), we can mention *Ecological Economics, Transportation Research Part A: Policy and Practice, Transport Policy, Journal of Transport Geography*, and *Pharmacoeconomics*. A large number of self-citations and strong connections with other journals can explain the position of *Transport Policy*, while a large number of self-citations and a small number of citations from other journals explain the appearance in the top three of other journals from Table 2.14 (especially *Ecological Economics*).

To conclude, we have detected the journals with a large number of citations of important journals, and journals where the observed rate of self-citation is dominant in the total level of citation. The results obtained can be used as a measure of the importance of scientific journals.

More information about this study is provided in Aleskerov et al. (2016c).

2.8 CONCLUSION

We have examined the influence of nodes in different real-life problems. Most of the studies are aimed at examining the influence of countries through different types of interactions: financial claims, migration, the trade of particular type of commodities, etc. An important feature of these problems is that countries are not homogeneous, hence, the same flow amount can be important for one country and non-critical for another. Therefore, there is a need to consider the individual characteristics of nodes. Since classical centrality measures do not take this feature into account, we applied the proposed influence measures (SRIC, LRIC, and interdependence indices) which provided more accurate results. The results are distinct from the results obtained by classical measures.

The proposed influence measures can be applied to other real networks. As an example, we described how these measures could be applied to the analysis of terrorist groups and citation networks. The detailed study of flows in these networks confirms that we can detect hidden influential elements.

Concluding Remarks

W E HAVE PRESENTED TWO new classes of centrality indices, SRIC and LRIC, which take into account the parameters of vertices and the group influence of vertices to a vertex. In the evaluation of these centrality indices, one can use several models dealing with how to aggregate the impact of different paths between two vertices, and the value of each path. We have illustrated these indices by their application to different real networks, from the network of foreign loans to citation networks.

However, the possibility of evaluating new features of influence in networks cannot be obtained for free. The main issue is that the complexity of evaluations increases exponentially. That is the reason why we restrict the lengths of paths by k (and k does not exceed 5, usually, it is assumed to be equal to 3–4), and the cardinality of critical groups by s (usually, s does not exceed 5).

Even these rather small numbers make it impossible to evaluate new indices for large networks, containing, say, more than 2,000–3,000 vertices, and not even speaking about millions of them. There are several ways to decrease the complexity of evaluations based on some observations of real systems. One such method is described in Shvydun (2020d). Other methods are based on parallel computations, on the use of supercomputers, etc.

Finally, we would like to emphasize that in our examples, given in Chapter 2, we made the evaluations of these new centrality indices for some reasonable (from our point of view) parameter values.

However, perhaps our readers would like to go much deeper and evaluate these indices for some other values of parameters. Some readers might wish to apply our SRIC and LRIC indices to other networks. For this, we provide the site with the software that allowed us to evaluate the SRIC and LRIC indices (https://github.com/SergSHV/slric).

References

Agneessens, F., Borgatti, S., and Everett, M. 2017. Geodesic Based Centrality: Unifying the Local and the Global. *Social Networks* 49: 12–26.

Akerman, A. and Seim, A. 2014. The Global Arms Trade Network 1950–2007. *Journal of Comparative Economics* 42, no. 3: 535–551.

Aleskerov, F. T. 2006. Power Indices Taking into Account Agents' Preferences. In *Mathematics and Democracy*, eds. B. Simeone and F. Pukelsheim, 1–18. Berlin, Germany: Springer.

Aleskerov, F. T., Andrievskaya, I. K., and Permjakova, E. E. 2014. *Key Borrowers Detected by the Intensities of Their Short-Range Interactions.* Working Papers by NRU Higher School of Economics. Series FE 'Financial Economics'. No. WP BRP 33/FE/2014.

Aleskerov, F. T., Meshcheryakova, N. G., Nikitina, A. A., and Shvydun, S. V. 2016a. *Key Borrower Detection by Long-Range Interactions.* Working papers by NRU Higher School of Economics. Series WP BRP 'Basic research program'. No. 56. arXiv:1807.10115.

Aleskerov, F. T., Meshcheryakova, N. G., Rezyapova, A. N., and Shvydun, S. V. 2016b. *Network Analysis of International Migration.* Working paper WP7/2016/06. Moscow, Russia: HSE Publishing House. arXiv:1806.06705.

Aleskerov, F., Badgaeva, D., Pislyakov, V., Sterligov, I., and Shvydun, S. 2016c. An Importance of Russian and International Economic Journals: A Network Approach. *Journal of the New Economic Association* 30, no. 2: 193–205.

Aleskerov, F., Andrievskaya, I., Nikitina A., and Shvydun, S. 2020a. *Key Borrowers Detected by the Intensities of Their Interactions. Handbook of Financial Econometrics, Mathematics, Statistics, and Machine Learning (In 4 Volumes)*, 355–389 World Scientific: Singapore Volume 1, Chapter 9

Aleskerov, F., Gavrilenkova, I., Shvydun, S., and Yakuba, V. 2020b. Power Distribution in the Networks of Terrorist Groups: 2001–2018. *Group Decision and Negotiation* 29, no. 3: 399–424.

Allen, F. and Gale, D. 2000. Financial Contagion. *Journal of Political Economy* 108, no. 1: 1–33.

Allen, F. and Babus, A. 2009, N. Y.: MacMillan. Networks in Finance. In *Network-Based Strategies and Competencies*, eds. P. Kleindorfer and J. Wind, 367–382.

Asal, V., Park, H. H., Rethemeyer, K., and Ackerman, G. 2015. With Friends Like These ... Why Terrorist Organizations Ally. *International Public Management Journal* 19, no. 1: 1–39.

Bacon, T. 2018. *Why Terrorist Groups Form International Alliances.* Philadelphia: University of Pennsylvania Press Philadelphia: University of Pennsylvania Press

Battiston, S., Glattfelder, J. B., Garlaschelli, D., Lillo, F., and Caldarelli, G. 2010. The Structure of Financial Networks. In *Network Science*, eds. E. Estrada, M. Fox, D. Higham, and G. L. Oppo, 131–163. London, UK: Springer.

Bavelas, A. 1950. Communication Patterns in Task-Oriented Groups. *The Journal of the Acoustical Society of America* 22: 725–730.

Basel Committee on Banking Supervision (BCBS). 2013. Global Systemically Important Banks: Updated Assessment Methodology and the Higher Loss Absorbency Requirement, Consultative Document. https://www.bis.org/publ/bcbs255.htm.

Bank of International Settlements. 2018. The Treatment of Large Exposures in the Basel Capital Standards – Executive Summary. https://www.bis.org/fsi/fsisummaries/largeexpos.pdf.

Bank of International Settlements. 2019a. Guidelines for Reporting the BIS International Banking Statistics. https://www.bis.org/statistics/bankstatsguide.pdf.

Bank of International Settlements. 2019b. BIS Statistical Bulletin, December 2019, Monetary and Economic Department. www.bis.org/statistics/bulletin1912.htm.

Bodvarsson, B., Simpson, N. B., and Sparber, C. 2015. Migration Theory. In *Handbook of Economics of International Migration. The Immigrants*, eds. B. P. Chiswick and P. W. Miller 1: North Holland: Elsevier3–51.

Bollen, J., Van De Sompel, H., Smith, J. A., and Luce, R. 2005. Toward Alternative Metrics of Journal Impact: A Comparison of Download and Citation Data. *Information Processing & Management* 41, no. 6: 1419–1440.

Bonacich, P. 1987. Power and Centrality: A Family of Measures. *American Journal of Sociology* 92, no. 5: 1170–1182.

Bonacich, P. and Lloyd, P. 2001. Eigenvector-Like Measures of Centrality for Asymmetric Relations. *Social Networks* 23: 191–201.

Bond, K. D. 2010. *Power, Identity, Credibility, & Cooperation: Examining the Development of Cooperative Arrangements among Violent Non-State Actors.* PhD diss., Penn State University, State College, PA.

Brin, S. and Page, L. 1998. The Anatomy of a Large-Scale Hypertextual Web Search Engine. *Computer Networks* 30: 107–117.

Stats.bis.org. 2020. Consolidated Banking Statistics. https://stats.bis.org/.

Chan-Lau, J. A. 2010. *The Global Financial Crisis and Its Impact on the Chilean Banking System,* IMF Working Paper No. 108 Washington: International Monetary Fund.

Chan-Lau, J. A. 2018. Systemic Centrality and Systemic Communities in Financial Networks. *Quantitative Finance and Economics* 2, no. 2: 468–496.

Comtrade.un.org. 2018. United Nations Statistics Division – Commodity Trade Statistics Database (COMTRADE). https://comtrade.un.org/db/mr/rfcommoditieslist.aspx.

Craft, C. and Smaldone, J. 2002. The Arms Trade and the Incidence of Political Violence in Sub-Saharan Africa, 1967–97. *Journal of Peace Research* 39, no. 6: 693–710.

Davis, K., D'Odorico, P., Laio, F., and Ridolfi, L. 2013. Global Spatio-Temporal Patterns in Human Migration: A Complex Network Perspective. *PLoS One* 8, no. 1: e53723.

De Meo, P., Ferrara, E., Fiumara, G., and Ricciardello, A. 2012. A Novel Measure of Edge Centrality in Social Networks. *Knowledge-Based Systems* 30: 136–150.

Estrada, E. and Rodriguez-Velazquez, J. A. 2005. Subgraph Centrality in Complex Networks. *Physical Review E* 71, no. 5: 056103.

Everett, M. G. and Borgatti, S. P. 1999. The Centrality of Groups and Classes. *Journal of Mathematical Sociology* 23, no. 3: 181–201.

Fagiolo, G. and Mastrorillo, M. 2012. The International-Migration Network. arXiv:1212.3852.

www.fao.org. 2018. SOFI 2018 – The State of Food Security and Nutrition in the World. http://www.fao.org/state-of-food-security-nutrition/en/.

Freeman, L. C. 1977. A Set of Measures of Centrality Based Upon Betweenness. *Sociometry* 40: 35–41.

Freeman, L. C. 1979. Centrality in Social Networks: Conceptual Clarification. *Social Networks* 1: 215–239.

Furfine, C. 2003. Interbank Exposures: Quantifying the Risk of Contagion. *Journal of Money, Credit and Banking* 35: 111–128.

Gantmacher, F. R. 2000. *The Theory of Matrices* New York. AMS Chelsea Publishing, No. 2.

Garlaschelli, D., Battiston, S., Castri, M., Servedio, V. D. P., and Caldarelli, G. 2005. The Scale-Free Topology of Market Investments. *Physica A* 350: 491–499.

Girvan, M. and Newman, M. E. J. 2002. Community Structure in Social and Biological Networks. *Proceedings of the National Academy of Sciences* 99, no. 12: 7821–7826.

Foodsecurityindex.eiu.com. 2018. The Global Food Security Index. http://foodsecurityindex.eiu.com/.

Global Hunger Index. 2018. *Global Hunger Index* – Official Website of the Peer-Reviewed Publication. http://ghi.ifpri.org/.

Goodman, L. A. and Kruskal, W. H. 1954. Measures of Association for Cross Classifications. *Journal of the American Statistical Association* 49, no. 268: 732–764.

Gubanov, D. A., Novikov, D. A., and Chkhartishvili, A. G. 2011. Informational Influence and Informational Control Models in Social Networks. *Automation and Remote Control* 72: 1557–1567.

Gubanov, D. A., Novikov, D. A., and Chkhartishvili, A. G. 2019. *Social Networks: Models of Information Influence, Control and Confrontation.* Cham, Switzerland: Springer International Publishing.

Guerrero-Bote, V. P. and Moya-Anegón, F. 2012. A Further Step Forward in Measuring Journals' Scientific Prestige: The SJR2 Indicator. *Journal of Informetrics* 6, no. 4: 674–688.

Hicks, D., Wouters, P., Waltman, L., de Rijcke, S., and Rafols, I. 2015. Bibliometrics: The Leiden Manifesto for Research Metrics. *Nature* 520, no. 7548: 429–431.

Horowitz, M. C. and Potter, P. B. K. 2014. Allying to Kill: Terrorist Intergroup Cooperation and the Consequences for Lethality. *Journal of Conflict Resolution* 58, no. 2: 199–225.

IMF. 2010. *Integrating Stability Assessments under the Financial Sector Assessment Program into Article IV Surveillance: Background Material.* Washington: International Monetary Fund

IMF/BIS/FSB. 2009. *Guidance to Assess the Systemic Importance of Financial Institutions, Markets and Instruments: Initial Considerations. Report to the G-20 Finance Ministers and Central Bank Governors.*

Iori, G., de Masi, G., Precup, O. V., Gabbi, G., and Caldarelli, G. 2008. A Network Analysis of the Italian Overnight Money Market. *Journal of Economic Dynamics and Control* 32, no. 1: 259–278.

Jackson, M. 2008. *Social and Economic Networks*. Princeton, NJ and Oxford, UK: Princeton University Press.

Kalinina, M., Kuskova, V., and Kuznetsov, V. 2018. Fourty Years of Network Science: Analysis of Journal Contribution to the Field. In Aachen: CEUR Workshop Proceedings *Supplementary Proceedings of the 7th International Conference on Analysis of Images, Social Networks and Texts (AIST-SUP 2018)*, eds. W. van der Aalst, V. Batagelj, G. Glavaš, D. I. Ignatov, M. Khachay, O. Koltsova, S. Kuznetsov, I. A. Lomazova, N. Loukachevitch, A. Napoli, A. Savchenko, and A. Panchenko, 155–160. Moscow, Russia.

Kang, C., Molinaro, C., Kraus, S., Shavitt, Y., and Subrahmanian, V. S. 2012. Diffusion Centrality in Social Networks. In *IEEE/ACM International Conference on Advances in Social Networks Analysis and Mining*, 558–664, Istanbul, Turkey.

Karmon, E. 2005. *Coalitions between Terrorist Organizations: Revolutionaries, Nationalists, and Islamists*. Boston, MA: Martinus Nijhoff.

Katz, L. 1953. A New Status Index Derived from Sociometric Index. *Psychometrika* 18: 39–43.

Kendall, M. 1970. *Rank Correlation Methods*, 4th Edition. London, UK: Griffin.

Kinsella, D. 2003. Changing Structure of the Arms Trade: A Social Network Analysis. Presented at the *Annual Meeting of the American Political Science Association*. Philadelphia, PA.

Kireyev, A. and Leonidov, A. 2015. Network Effects of International Shocks and Spillovers. *Networks and Spatial Economics* 18: 805–836.

Kleinberg, J. M. 1999. Authoritative Sources in a Hyperlinked Environment. *Journal of the ACM* 46, no. 5: 604–632.

Korniyenko, Y., Patnam, M., Porter, M., and del Rio-Chanon, R. 2018. Evolution of the Global Financial Network and Contagion: A New Approach. IMF Working Paper No. 18/113. https://ssrn.com/abstract=3221170.

Latora, V. and Marchiori, M. 2004. How the Science of Complex Networks Can Help Developing Strategies against Terrorism. *Chaos, Solitons and Fractals* 20: 69–75.

Lee, E. 1966. A Theory of Migration. *Demography* 3, no. 1: 47–57.

Leydesdorff, L. and Bornmann, L. 2011. Integrated Impact Indicators Compared with Impact Factors: An Alternative Research Design with Policy Implications. *Journal of the American Society for Information Science and Technology (JASIST)* 62, no. 11: 2133–2146.

Mainas, E. D. 2012. The Analysis of Criminal and Terrorist Organisations as Social Network Structures: A Quasi-Experimental Study. *International Journal of Police Science & Management* 14, no. 3: 264–282.

Mancheri, N. 2015. China and Its Neighbors: Trade Leverage, Interdependence and Conflict. *Contemporary East Asia Studies* 4, no. 1: 75–94.

Martin, P., Mayer, T., and Thoenig, M. 2005. Make Trade not War? CEPREMAP Working Papers (*Docweb*) 0515, CEPREMAP.

Mavroforakis, C., Garcia-Lebron, R., Koutis, I., and Terzi, E. 2015. Spanning Edge Centrality: Large-Scale Computation and Applications. In *Proceedings of the 24th International Conference on World Wide Web*, 732–742.

Medina, R. M. 2014. Social Network Analysis: A Case Study of the Islamist Terrorist Network. *Security Journal* 27: 97–121.

Meshcheryakova, N. 2020. Network Analysis of Bilateral Trade Data under Asymmetry. In *2020 IEEE/ACM International Conference on Advances in Social Networks Analysis and Mining (ASONAM)*, 379–383. doi: 10.1109/ASONAM49781.2020.9381408.

Meshcheryakova, N. and Shvydun, S. 2018. Power in Network Structures Based on Simulations. In *Complex Networks & Their Applications VI. COMPLEX NETWORKS 2017. Studies in Computational Intelligence* 689, eds. C. Cherifi, H. Cherifi, M. Karsai, M. Musolesi, 1028–1038. Cham, Switzerland: Springer.

Myerson, R. B. 1977. Graphs and Cooperation Games. *Mathematics of Operations Research* 2: 225–229.

National Consortium for the Study of Terrorism and Responses to Terrorism (START). 2019. *Global Terrorism Database*. https://www.start.umd.edu/gtd.

Newman, M. E. J. 2003. The Structure and Function of Complex Networks. *SIAM Review* 45, no. 2: 167–256.

Newman, M. E. J. 2010. *Networks: An Introduction*. Oxford, UK: Oxford University Press.

Nicholson, C. D., Barker, K. and Ramirez-Marquez, J. E. 2016. Flow-Based Vulnerability Measures for Network Component Importance: Experimentation with Preparedness Planning. *Reliability Engineering & System Safety* 145: 62–73.

OECD Bilateral Trade in Goods by Industry and End-use (BTDIxE). 2020. *ISIC Rev.* 4, https://stats.oecd.org/.

Ortiz-Gaona, R., Postigo-Boix, M., and Melus, J. 2016. Centrality Metrics and Line-Graph to Measure the Importance of Links in Online Social Networks. *International Journal of New Technology and Research* 2, no. 12: 20–26.

Ouellet, M., Bouchard, M., and Hart, M. 2017. Criminal Collaboration and Risk: The Drivers of Al Qaeda's Network Structure before and after 9/11. *Social Networks* 51: 171–177.

Page, L. B. and Perry, J. E. 1994. Reliability Polynomials and Link Importance in Net-Works. *IEEE Transactions on Reliability* 43, no. 1: 51–58.

Pedahzur, A. and Perliger, A. 2006. The Changing Nature of Suicide Attacks: A Social Network Perspective. *Social Forces* 84, no. 4: 1987–2008.

Perliger, A. 2014. Terrorist Networks' Productivity and Durability: A Comparative Multi-Level Analysis. *Perspectives on Terrorism* 8, no. 4: 36–52.

Piraveenan, M., Prokopenko, M., and Hossain, L. 2013. Percolation Centrality: Quantifying Graph-Theoretic Impact of Nodes during Percolation in Networks. *PLoS ONE* 8, no. 1: e53095.

Ravenstein, E. G. 1889. The Laws of Migration. *Journal of the Royal Statistical Society* 52, no. 2: 241–305.

Rochat, Y. 2009. *Closeness Centrality Extended to Unconnected Graphs: The Harmonic Centrality Index.* Proceedings of the 6th Applications of Social Network Analysis Conference, Zurich, Switzerland ASNA

San Francisco Declaration on Research Assessment. 2013. http://www.ascb.org/dora/.

Schoch, D. 2018. Centrality without Indices: Partial Rankings and Rank Probabilities in Networks. *Social Networks* 54: 50–60.

Seglen, P. O. 1997. Why the Impact Factor of Journals Should Not Be Used for Evaluating Research. *British Medical Journal* 314, no. 7079: 498–502.

Shapley, L. S. and Shubik, M. 1954. A Method for Evaluating the Distribution of Power in a Committee System. *American Political Science Review* 48: 787–792.

Shimbel, A. 1953. Structural Parameters of Communication Networks. *Bulletin of Mathematical Biology* 15: 501–507.

Shvydun, S. 2020a. Dynamic Analysis of the Global Financial Network. In *2020 IEEE/ACM International Conference on Advances in Social*

Networks Analysis and Mining (ASONAM), 374–378. doi: 10.1109/ASONAM49781.2020.9381345.

Shvydun, S. 2020b. Influence of Countries in the Global Arms Transfers Network: 1950–2018. In *Complex Networks and Their Applications VIII. COMPLEX NETWORKS 2019. Studies in Computational Intelligence* 882, eds. H. Cherifi, S. Gaito, J. Mendes, E. Moro, and L. Rocha, 736–748. Cham, Switzerland: Springer. doi: 10.1007/978-3-030-36683-4_59.

Shvydun, S. 2020c. Power of Nodes Based on Their Interdependence. In *Complex Networks XI. Springer Proceedings in Complexity*, eds. H. Barbosa, J. Gomez-Gardenes, B. Gonçalves, G. Mangioni, R. Menezes, and M. Oliveira, 70–82. Cham, Swirzerland: Springer. doi: 10.1007/978-3-030-40943-2_7.

Shvydun, S. 2020d. Computational Complexity of SRIC and LRIC Indices. In *Network Algorithms, Data Mining, and Applications. NET 2018. Springer Proceedings in Mathematics & Statistics 315*, eds. I. Bychkov, I. V. Kalyagin, P. Pardalos, and O. Prokopyev, 49–70. Cham, Switzerland: Springer. doi: 10.1007/978-3-030-37157-9_4.

Simmel, V., Holtom, P., and Bromley, M. 2012. *Measuring International Arms Transfers*. Stockholm International Peace Research Institute, Stockholm International Peace Research Institute (SIPRI).

Simon, G. A. 2012. *Food Security: Definition, Four dimensions, History*. Rome, Italy: University of Roma Tre.

SIPRI. 2019. The SIPRI Arms Transfers Database. https://www.sipri.org/databases/armstransfers.

Sislin, J. 1994. Arms as Influence: The Determinants of Successful Influence. *The Journal of Conflict Resolution* 38, no. 4: 665–689.

Sjaastad, L. 1962. The Costs and Returns of Human Migration. *Journal of Political Economy* 70: 80–93.

Smith, A. and Garnier, M. 1838. *An Inquiry into the Nature and Causes of the Wealth of Nations*. Edinburgh, UK: Thomas Nelson.

Tanious, M. 2019. The Impact of Economic Interdependence on the Probability of Conflict between States. *Review of Economics and Political Science* 4, no. 1: 38–53. doi: 10.1108/REPS-10-2018-010.

Thurner, P. W., Schmid, C. S., Cranmer, S. J., and Kauermann, G. 2019. Network Interdependencies and the Evolution of the International Arms Trade. *Journal of Conflict Resolution* 63, no. 7: 1736–1764.

Tinbergen, J. 1962. *Shaping the World Economy: Suggestions for an International Economic Policy. HD82 T54*. New York, NY: The Twentieth Century Fund.

Tranos, E., Gheasi, M., and Nijkamp, P. 2015. International Migration: A Gobal Complex Network. *Environment and Planning* 42, no. 1: 4–22.

United Nations. 2015. New York: United Nations *Department of Economic and Social Affairs, Population Division. International Migration Flows to and from Selected Countries: The 2015 Revision* (POP/DB/MIG/Flow/Rev.2015).

Waltman, L. 2016. A Review of the Literature on Citation Impact Indicators. *Journal of Informetrics* 10, no. 2: 365–391. doi: 10.1016/j.joi.2016.02.007.

Waltman, L. and Yan, E. 2014. PageRank-Related Methods for Analyzing Citation Networks. In *Measuring Scholarly Impact: Methods and Practice*, eds. Y. Ding, R. Rousseau, and D. Wolfram, 83–100. doi: 10.1007/978-3-319-10377-8_4.

Wits.worldbank.org. 2018. WITS - About WITS. https://wits.worldbank.org/about_wits.html.

Zipf, G. K. 1946. The P1 P2/d Hypothesis: On the Intercity Movement of Persons. *American Sociological Association* 11, no. 6: 677–686.

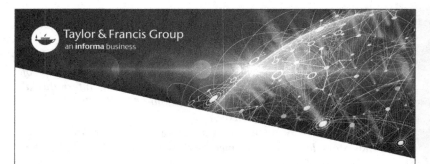

Printed in the United States
by Baker & Taylor Publisher Services